T0212020

Colonization of the Inner Planet

This book explores the conquest, predation, and management of human bodies and emotions by the growing capitalist digital world. It seeks to understand the debate between various forms of the individual, subject, actor, and agent to emerge a social theory vision for the 21st century.

The book moves beyond the colonization of the physical world to examine the process of colonization of humans. It focuses on the communication humans have with the world to understand how this impacts their sensibilities. This communication is influenced by technological innovations that enable a process of systematic colonization of human beings as bodies/emotions. This book explores a social theory which will allow us to understand this redefinition of the individual. This enables us to uncover connections between the colonization of the "inner planet" that is the human society, the dialectic of the person, and the politics of their sensibilities. This is explored through the tensions that arise between the forms a person assumes in unequal and diverse cultural contexts and the emotions behind those cultural differences.

The book will appeal to academics and postgraduate students of sociology, philosophy, and anthropology, as well as psychologists, organizational specialists, linguists, ethnographers, historians, political scientists, administrators, and professionals affiliated with NGOs.

Adrian Scribano is Director of the Centre for Sociological Research and Studies and Principal Researcher at the National Scientific and Technological Research Council, Argentina. He is also the Director of the Latin American Journal of Studies on Bodies, Emotions and Society and the Study Group on Sociology of Emotions and Bodies, in the Gino Germani Research Institute, Faculty of Social Sciences, University of Buenos Aires.

Routledge Research in the Anthropocene

The *Routledge Research in the Anthropocene Series* offers the first forum for original and innovative research on the epoch and events of the Anthropocene. Titles within the series are empirically and/or theoretically informed and explore a range of dynamic, captivating and highly relevant topics, drawing across the humanities and social sciences in an avowedly interdisciplinary perspective. This series will encourage new theoretical perspectives and highlight ground-breaking interdisciplinary research that reflects the dynamism and vibrancy of current work in this field. The series is aimed at upper-level undergraduates, researchers and research students as well as academics and policy-makers.

For more information about this series, please visit https://www.routledge.com/ Routledge-Research-in-the-Anthropocene/book-series/RRA01

Colonization of the Inner Planet

21st Century Social Theory from the
Politics of Sensibilities

Adrian Scribano

Routledge
Taylor & Francis Group

LONDON AND NEW YORK

First published 2022
by Routledge
2 Park Square, Milton Park, Abingdon, Oxon OX14 4RN

and by Routledge
605 Third Avenue, New York, NY 10158

Routledge is an imprint of the Taylor & Francis Group, an informa business

© 2022 Adrian Scribano

British Library Cataloguing-in-Publication Data
A catalogue record for this book is available from the British Library

Library of Congress Cataloging-in-Publication Data
A catalog record has been requested for this book

ISBN: 978-0-367-77287-1 (hbk)
ISBN: 978-0-367-77290-1 (pbk)
ISBN: 978-1-003-17066-2 (ebk)

Typeset in Times New Roman
by SPi Global, India

To Angie

Contents

Tables

Acknowledgements

I want to thank Faye Leerink for trusting on this project. I also want to thank Nonita Saha for the editorial assistance. Finally, this book would not be the same without the invaluable assistance of Majid Yar, and the enormous collaboration of Aldana Boragnio.

I would like to especially thank my whole family, especially María de la Paz, María Belén, Maia Pia, and Malena for their love, support, and affection.

Introduction

Colonization of the inner planet and social theory

I.1 Introduction

> Nan-in, a Japanese master of the Meiji era, received a visit from a university professor who had come to him to enquire about Zen. Nan-in served tea. He filled his guest's cup and then continued to pour. The professor watched his tea overflowing until he could contain himself no longer. The cup's full to the brim', he exclaimed, 'it can't take any more!' 'Like this cup', replied Nan-in, 'you're full to overflowing with your opinions and conjectures. How can I explain Zen to you unless you empty your cup?' We treat the inner dimension of our experience like the professor.
>
> (Melucci, 1996: 74)

In this way, Alberto Melucci begins Chapter Four of his book *Playing Self* – entitled Inner Planet. This book serves as homage to Alberto Melucci, and while the text presented here does not reflect his views regarding the inner planet, it does obviously follow several of his intuitions.

This is a book about one of the most colossal colonial enterprises related to the human being: the exploration, conquest, occupation, predation, and management of his own body/emotion. The book does not have a historical character, but rather offers a theoretical examination of these developments.

From the 15th to the 20th century, we have lived the colonization of the world in and through the expansion of communications and commerce: the planet became a single world. Environmental assets, living beings, and all forms of energy were owned and predated by powerful states and corporations. The Anthropocene crisis of the Capitalocene has revealed the consolidation of a process of systematic colonization of human beings as bodies/emotions. This process is what we can call colonization of the inner planet.

The 21st century is waiting for a social theory that takes up the challenges of the new societies that have emerged in its first 20 years. Facebook is 15 years old; Tesla was founded in 2003; Siri for iPhone was made available to the public in 2010; Amazon was founded 1994, Google in 1998, and Alibaba in 1999; these are just some of the examples that have "changed" the world by shrinking it and super-connecting it. Within this framework, violence, inequality, depredation, the climate crisis, femicide, as well as post-speciesist dialogues, diverse gender perspectives, views of "good living" (*buen vivir*, *Sumak Kawsay*), and the communitarian perspectives complete, at least partially, a context of profound transformations. The proposal made here is built out of all kinds of romanticism, miserabilism, populism, essentialism, and dogmatism and is produced from an intersectional, non-speciesist, and non-colonial perspective.

The expansion of capitalism on a global scale has produced a process of colonization of the inner planet. This can be described synthetically as a process of commodification of the senses. Said commodification is the result of the application of scientific and technological knowledge for the design, production, and reproduction of the different ways of hearing, tasting, touching, smelling, and seeing which we have today in the framework of Society 4.0.

From this perspective, an occupation by the large global companies of the conditions of possibility of feeling is verified. Colonization in this century is no longer circumnavigating the seas but, rather, returning merchandise to the vehicles of communication that human beings have with the world. Therefore, in parallel to the depredation of natural resources of the planet, there is also a dispossession of bodily energies, and a reshaping of how impressions, perceptions, and sensations are connected.

This book aims to elaborate a social theory that allows us to understand the situation outlined above, that is the redefinition of the individual in the light of the current centrality of the politic of sensibilities. It is in this context that the present book tries to show how in the dialectic between individual, subject, actor, agent, and author emerges a vision of the social person that allows us to understand the central features of a Social Theory for the 21st century. This tension between the forms that the person assumes is inscribed in the unequal and diverse contexts of a plurality of geocultures and geopolitics of emotions that constitute different politics of sensibilities. In this sense, the book elaborates a social theory that can make visible the connections between the colonization of the inner planet, the dialectic of the person, and the politics of sensibilities.

This Introduction seeks: (a) to introduce the reader into the area of the bodies/emotions that the book addresses, (b) to express some theoretical and epistemic axes as a substrate of this approach, and (c) to synthesize the content of the chapters of the book.

I.2 Sociology of the bodies/emotions: A perspective

What we know about the world we know by and through our bodies. Perceptions, sensations, and emotions build a tripod that allows us to understand where sensibilities are founded. Social agents know the world through their bodies. Thus, a set of impressions impact on the ways subjects "exchange" with the socio-environmental context. Such impressions of objects, phenomenon, processes, and other agents structure the perceptions that subjects accumulate and reproduce. Perception, from this perspective, constitutes a naturalized way of organizing the set of impressions that are given in an agent.

This weaving of impressions configures the sensations that "produce" what can be called the "internal" and "external" world: social, subjective, and "natural" worlds. Such configurations are formed in a dialectic tension between impressions, perceptions, and their results, that give sensations the "meaning" of a surplus or excess. Therefore, it puts them closer and beyond such a dialectic.

Sensations, as a result, and as the antecedent of perceptions, locate emotions as an effect of the processes of adjudication and correspondence between perceptions and sensations. Emotions, understood as the consequences of sensations, can be seen as a puzzle that becomes action and effect of feeling something or feeling oneself. Emotions are rooted in the "state of feeling" the world that allows the sustaining of perceptions. These are associated with socially constructed forms of sensations. At the same time, organic and social senses also enable what seems unique and unrepeatable as are individual sensations, and elaborate on the "unperceived work" of incorporating social elements turned into emotions.[1]

What we know about the world, we know through our bodies, what we do is what we see, and what we see is how we divide the world. In this "here-now" the devices for the regulation of sensations are installed. By such devices the social world is both apprehended and narrated.

Sensations are distributed according to the specific forms of corporal capital. Corporal capital consists of the living conditions of individuals located in the individual body, subjective body, and social body.

The tension between individual, subjective, and social bodies is one of the keys that will allow a deeper understanding of the connections between geometries of the bodies and grammars of action, which are part of the neocolonial domination in Latin America and the Global South. The aforementioned tension makes more sense when joining the perspective from bodies with the view from sensations.

A privileged form of connection between collective action and social fantasies and phantoms is constituted by the acceptance of the fact that the body is the locus of conflict and order. It is the place and "*topos*" of conflict where (much of) the logic of contemporary antagonisms passes through. From this point of view, we can observe the formation of a political economy of morality, that is, forms of sensibilities, practices, and representations that put domination into words.

In this context, we understand that social bearability mechanisms are structured around a set of practices-made-body, orientated to a systematic avoidance of social conflict. The processes of displacement of the consequences of antagonism are presented as specular scenarios unpinned (disembedded) in space and time. These allow the acceptance that social life "is-done" as-always-so by the individual and society as a whole. Associated with this, devices for the regulation of sensations consist of processes of selection, classification, and the elaboration of socially determined and distributed perceptions. Regulation implies some sort of tension between the senses, perceptions, and feelings that organize the special ways of "seeing oneself-in-the-world" and "appreciation-in-the-world" that classes and subjects possess.

Chains and cognitive-affective schemas that connect (and disconnect) social practices as narratives and worldviews made flesh are the processes that we characterize as ideological. The identified mechanisms and devices are a practical and procedural hinge where crossings between emotions, bodies, and stories instantiate. Systems' social bearability mechanisms do not operate either directly or explicitly as "attempted control," nor "deeply" as processes of focal points of persuasion. These mechanisms operate "almost unnoticed" in the porosity of custom, in the frames of common sense, through the construction of sensations that seem the most "intimate" and "unique" that every individual possesses as a social agent.

As we have argued already, among them there are two that acquire relevance from a sociological point of view: social fantasies and phantoms. One is the reverse of the other; they both refer to the systematic denial of social conflict. While fantasies occlude conflict, they invert (and consecrate) the place of particular elements posing them as universal, and they avoid the inclusion of subjects in the fantasized terrains; at the same time phantoms repeat the conflictual loss, remember the weight of the defeat, and devalue the possibility of counter-action in the face loss and failure. Social Fantasies and Phantoms never have closure; they are contingent but always operate as and are turned into practices. Consequently, " feeling practices " involve such a "state of affairs" that combines impressions and perceptions becoming a vehicle where past and present come together in a particular experience. Thus, "feeling practices" are forged into and they actualize/incarnate in concrete processes that include the set of sensibilities that constitute politic of the emotions.

As an ideological mechanism, one of the social results of fantasy is that it seems to impose nothing (not rules, not classifying provisions, etc.), "only" tells us how to classify, how to construct rules. An important feature of social fantasies is that they produce an operation of acceptance of what they seem to suppress; install what they want to de-install.

The efficiency of social fantasies' mechanisms is due, in part, to their ability to conceal antagonisms. Fantasies operate by hiding conflicts, by making them visible without their inherent antagonism. Another feature of the mechanisms of "social fantasy" is the paradoxical situation of the subject that is held (subjected) by them. The subject who lives the proposed and socially accepted fantasy does not need and cannot get out of the same scenario. He cannot "do" fantasy on pain of ceasing to be. This is because in social reality there are mechanisms of order stabilization that have two facets. The fantasy circulates and becomes effective in a kind of blind spot of common sense, or that which one ends up accepting because it is obvious. It builds the direct possibility of taking as natural and naturalizable the possibility of feeling the conflictual situation as something nondisruptive. Another characteristic that must be taken into account when analysing fantasies is their heteronomy, that is, what, as ideological devices, always constitute the margin of the autonomy of subjects. Finally, fantasy does not appear in explicit texts. It does not have fixed content. It cannot be determined. It should always be exposed as contradicting reality.

Phantoms have another regime of action and involve different processes than fantasies. The phantoms repeat the absence of conflict, remember the weight of defeat, and devalue the possibility of counteraction in the face of lack and failure. Phantoms are configured as updating the real that indicates what has been lost.

The practical structure of social phantoms is something approaching a melancholy state: we know that something has been lost, but we do not know what. They emerge from the embodiment of impotence and pain. Social phantoms are "incorporated" into the mode of the spectrum. They are backdrops that paint, "build," and are always "on hand" to operate as horizons of "necessary" and ineluctable failure. They are the knot where the future is tied to the past.

Social phantoms involve the conjunction of the logic of threat, hostage, and an experiential abduction "incorporated," made "bone." The phantom always announces its return, its imminent apparition before the behavioural "deviation." It takes the bodies and the emotions like hostages of fulfilment and the reproduction of order and crises. The autonomous practices of the subjects are kidnapped (punctual, multiple, contingent, and iterative), as hostages whose rescue lies in social reproduction.

Phantoms close the constitutive breaches of societal logic. They appear as a linkage of the multiple fragments of the present, strengthening the repetition of the past as an explanatory solution. Social phantoms have various operative modes that make them efficient and effective; one of these modes is their ability to serve to give reasons as to why things happen. Their place in social sensibilities is to give quick and systematic explanations for what cannot be understood without their presence. The phantoms appear as effects that allow understanding of their causes, cancelling the possibility of wondering about the genesis of social perceptions.

In the aforementioned context, it is very important to clarify the tensional, dialectical, and helical constitution of bodies as the nodal point of our social analysis. The intersections between impressions, perceptions, sensations, and emotions constitute a fundamental starting point for analysing the structuration social processes that are concretized in, through, and by the body.

To reconstruct perceptions of the body implies at least two intertwined paths: (a) crossings and ruptures between body-individual, body-subjective, and body-social; and (b) articulations and connections between body-image, skin, and movement.

The *first path* lies in visiting the distances and entanglements between body-individual, body-social, and body-subjective. This entails underlining the connections between the experience of the body as an organism, the experience of the body as a reflective act, and the practice of the body as a social construction. A body-individual refers to phylogenetic logic, to the articulation between the organic and the environment; a body-subjective takes off from self-reflection, in the sense of the "I" as a centre of gravity through which multiple subjectivities are formed; and finally, a social body is the social made flesh, as it were (*sensu* Bourdieu).

These three basic body practices organize and are organized by logics of regulation of the senses. Gradual progress and constant metamorphosis of sensibilities are ways of appropriating the body's energy, the connections between diverse body-sensations; taken together, this is one of the pillars of domination and also of autonomy. This kind of ontological redescription demands a discussion of the differences between body and social energies, but this cannot be discussed here. To maintain the "state of things" assigned as body individual, the body-energy must become an object of both production and consumption. Social energy that is presented through the social body is based on body energy. The power to plan, carry out, and resolve the consequences of the actions of agents is what constitutes social energy.

The *second* possible path is to draw and reconstruct what we know about the body about the methods of getting to know it, that is, as body-image, body-skin, and body-movement. These three ways of inscribing the corporal in a narrative offer a reconstructive analysis of how we can see the corporal from its impacts on sociability, sensibility, and experiences of everyday life as social phenomena.

At first, body-image is an indicator of the process of "I see myself as I feel that others see me" body-skin indicates the process of how one "naturally-feels" the world; and body-movement is the inscription of the body in a field of possible actions. These three ways to reconstruct body experiences may be seen as paths for the analyses and interpretations of how body forms appear socially. Tension and process between social parts of the body, the body there, and posture, as a significant social structure, elaborate the textuality of the body-image that every agent must build and administrate.

The senses appear to be natural, but they are also the results of a social process. It is in this process of social construction that the body-skin is built. Consequently, sociability and social sensibilities get entrenched as the "natural" way to "feel the world."

Body-movement is a mediation of both the power and impotence of the body and social energies. There are ways bodies can act, which is to say that they can act depending on social energies and social inertia. In other words, certain forms of action embody the social geometries of displacement as well as social inertia. In this process, seeing, smelling, touching, hearing, and tasting coalesce into possible sociability, indicating social devices for the regulation of sensations.

Sensations, as a result, and antecedent of perceptions, give way to emotions which can be seen as the manifestation of the action and effect of feelings. They are rooted in the states of feeling the world that build perceptions associated with socially constructed forms of sensations. At the same time, organic and social senses also permit mobilization of that which seems unique and unrepeatable (such as individual sensations), and they carry out the "unnoticed work" of the incorporation of the social – which has become the emotion.

Emotions are practices that transform the world that, based on a biography of sensations, challenge the person, producing recognition of bodily and affective states that involve them in all the modalities of their geometry. Emotions are affective cognitive tendencies that: (a) imply a movement, an activity, and a modification of time/space; (b) serve as maps to recognize the interaction situations; and (c) allow the management of the effects of said interactions.

Consequently, the politic of bodies (i.e., the strategies that society accepts to offer a response to the social availability of individuals) is a chapter – and not the least important chapter – in the instruction manual of power. These strategies are tied and "strengthened" by the politic of emotions that tend to regulate the construction of social sensibility. Politic of emotions require regulating and making bearable the conditions under which social order is produced and reproduced. In this context, we understand that social bearability mechanisms are structured around a set of practices that have become embodied and that are oriented towards a systematic avoidance of social conflict.

The forms of sociability and experience are strained and twisted as if contained in a Moebius strip along with the sensibilities that arise from regulatory devices and the aforementioned mechanisms. The need to distinguish and link the possible relations between sociability, experience, and social sensibilities becomes crucial at this point. Sociability is a way of expressing how agents live and coexist interactively. Experience is a way of expressing the meaning gained while being in physical proximity with others, as a result of experiencing the dialogue between the individual body, the social body, and the subjective body, on the one hand, and the natural appropriation of bodily and social energies on the other. For the body to be able to reproduce experience and sociability, bodily energy must be an object of production and consumption. Such energy can be understood as the necessary force to preserve the state of "natural" affairs in systemic functioning. At the same time, the social energy shown through the social body is based on the bodily energy and refers to the allocation processes of such energy as the basis of the conditions of movement and action.

Thus, sensations are distributed according to the specific forms of bodily capital; and the body's impact on sociability and experience shows a distinction between the body image, the body skin, and the body movement. The forms of sociability and experience are intertwined and twisted as if in a Moebius strip with the sensibilities that arise as a result of mechanisms of regulating sensation.

Social sensibilities are continually updating the emotional schemes that arise from the accepted and acceptable norms of sensations. They are just a little long or short of the interrelationships between sociability and experience. Sensibilities are shaped and reshaped by contingent and structural overlaps of diverse forms of connection/disconnection among various ways of producing and reproducing the politic of the body and the emotions.

The politics of sensibilities are understood as the set of cognitive-affective social practices tending to the production, management and reproduction of horizons of action, disposition and cognition. These horizons refer to: (1) the organization of daily life (day-to-day, vigil/sleep, food/abstinence, etc.); (2) information to sort preferences and values (adequate/inadequate, acceptable/unacceptable, bearable/unbearable); and (3) parameters for time/space management (displacement/location, walls/bridges; enjoyment). Interstitial practices nest in the inadvertent folds of the naturalized, naturalizing surface of the politics of the bodies and the emotions of neocolonial religion. They are disruptions in the context of normativity.

In this context, three concepts become relevant: "practices of wanting," "feeling practices", and "interstitial practices." Feeling Practice are those practices that involve heterogeneous sets of relationships between sensations and emotions. Interstitial practices are those social bondages that proceed to break the political economy of the moral – which structures sensibilities. Practices of wanting involve the possible connections between hope, love, and enjoyment, and are social relations that link us to "doing with" the other. Associations between the aforementioned practices, social bearability mechanisms, and devices to regulate sensations might allow us to better understand the state of social sensibilities.

The bar (/) that we inscribe between bodies/emotions implies a sociologized allusion to its uses in psychoanalysis with the intention of showing the separation/ union, distance/proximity, and possibility/impossibility between objects/discourses that we grant to what has been thought as separate, specific, and distant disciplinary subfields.

The gesture of retaking psychoanalysis as an "ally" of the interpretation of the social (long-standing in social theory) is plotted in our gaze/retranslation of critical theory (Marcuse, Fromm, etc.), from critical hermeneutics (Ricoeur) and ideological criticism (Žižek). A reflection on the questions that the aforementioned "alliance" opens up exceeds the intentions of this Introduction, and we leave it for consideration elsewhere, but it seems appropriate to rethink the current separation between the aforementioned traditions in numerous contemporary studies as impossible denial and/or constitutive suppression, which leaves us at the door of a set of theoretical analyses in favour of an imagined impossibility. We precisely propose to see in this bar the moment of the beginning of writing about, of and with the bodies/emotions as narratives that imply and "intersect" them: (a) as a space from where rather than losing, the differences are recovered as part of a Moebesian band and (b) as a designating operator of the spiral effect implied by the "beginning/step/end" relationship structured both in bodies and in emotions.

In the development of the aforementioned inquiries, we argue that in the elaboration of social theories of the South an epistemic, theoretical and methodological configuration must be produced that can be understood in the articulations and disarticulations, in the connections and disconnections, in the proximities and distances that provide: a logic of the seminal explanation of the structures; a Moebesian, spiral, reticular and dialectical constitution of social processes- phenomena; and chromatic analogies for the understanding of social practices and the construction of substantive information from an approach that takes sensations as a starting point to listen to multiple voices.[2]

I.3 Theoretical approach, epistemic assumptions, and values

A fundamental starting point for the task that this book intends to carry out is to make it very clear that it must be understood in a post-intersectional, post-speciesist, and postcolonial framework of understanding.

I.3.1 Postcoloniality and social theory of South (STS)

One of the strongest pieces of evidences of the changes that we are currently experiencing is the increasing presence of coloniality, and hence the urgency of retaking an analysis that can approach the phenomenon critically.

In the decade of the 1980s in English universities, a series of research units or academic departments were constituted around a multidisciplinary programme in social sciences that came to be called cultural studies. These groups included investigations and enquiries of diverse disciplinary origins. Among others, communication, literary criticism, sociology, anthropology, and gender approaches can be mentioned. These collective efforts converged on the need to analyse the forms of domination in the central societies, crossed by difference and the inequalities of class, gender, ethnicity, religion, and age. Within the framework of the academic development of cultural studies and its multiplication in foreign universities, especially in the United States, research programmes with their own and sometimes independent characteristics were organized, such as multicultural approaches and subaltern studies.

One way to understand the current content of postcolonial options is to explore their "internal history," and one way to do it is to connect the interpretative schemes from those mentioned as "founding fathers" among whom, and just to mention some, we must indicate Marcus Garvey, C.L.R. James, Albert Memmi, and Franz Fanon.

If the processes of institutionalization and/or formalization of groups/visions/approaches are taken as an indicator, two that should be mentioned are those of subaltern studies and the modernity/coloniality group (Quijano, 1993, 1988, 2000; Quijano and Wallerstein, 1992; Dussel, 1993, 1996, 2000; Mignolo, 1995, 2007; Lander; 2000, 2002; Escobar, 2007). The first visibly connects with Guha's proposals, and the second includes Latin American authors who follow Guha's ideas (Guha, 1963; Guha and Spivak, 2002).

On the other hand, the processes of reflection, conceptualization, and dialogue led to the reconfiguration of postcolonial studies, in decolonial approaches, and perspectives from the standpoint of subalternity. The coexistence, proximity, and distance of these viewpoints made the analysis more complex and limited the theoretical identities of each one (Coronil, 1996; Castro-Gómez and Mendieta, 1998). It is at the beginning of the 2000s (with a deep and abiding connection with previous historical processes) that the different ways of approaching "good living" emerged. This revitalization and redefinition added to the philosophical instruments and social theoretical tools the ancestral views about life (De Sousa Santos, 2018).

Concomitant with the movement mentioned, feminist philosophy and theory complete, from a perspective of coloniality, racialization, and patriarchality, a set of perspectives "close" to postcolonial studies. It is not possible to address here properly this perspective, which has its own internal history.

In the tension of LGTBIQ, decolonial feminist, and racialization studies, we can also capture another space of proximity with postcolonial studies, as I want to show here (Lugones, 2002, 2003; Anzaldua, 2012; Tyagi, 2014).

In the aforementioned context, in parallel in some places and others dependent on cultural studies, the so-called "postcolonial studies" were born and consolidated towards the end of the 1980s. The latter was presented towards the end of the 1990s as perhaps the most "radical" alternative to the canons of work in the social sciences. Defined from a multidisciplinary platform, where the narrative, historical and aesthetic-cultural is prevalent, postcolonial studies are tied around a group of simple common assumptions, but of great impact on the socially accepted discursive form of the social sciences.

Among the most important assumptions we can mention: (a) the criticism of European reason as the centre of scientific work; (b) the possibility and the need to configure narratives from the "margins" crossed by class, gender and ethnic positions; (c) the urgency to rethink particular realities from the geopolitics of knowledge; and (d) the challenge of marking the terms of domination through the relationship between academia, science and everyday life.

It is necessary to clarify that the construction of theory is understood here as the result of a practice that weaves together and crosses five spheres of knowledge of the social. In this direction, the epistemological, the methodological, the ontological, the critical-instrumental, and the substantive theoretical are all features of the elaboration of a theory. On the other hand, what we present here must be inscribed in the post-empiricist context of the social sciences in particular, and within the framework of the characteristics of complexity from which science, in general, is produced today.

Synthetically, it is important to underline that the proposals made here are the result of a set of conceptual reflections, the results of our empirical investigations and the inscription in a particular tradition of the social sciences. In this context, what follows is a tribute to the intersections between *conflict theory* and *collective action, sociology of bodies and emotions*, and *ideological criticism*. These are crossings that in turn must be inscribed in the frameworks produced by critical hermeneutics, the critical theory of the Frankfurt school and the dialectical critical realism of Roy Bhaskar as the expression of a way of understanding the ideas of Karl Marx. It is dialectical critical realism that allows us to link the aforementioned approaches with our demand for "situated" thinking while at the same time instantiating Marx as the axis of the aforementioned articulations between what we have called spheres of a theory, as an "updated" theoretical practice. Expressed in another way, within the framework of what has been outlined above, sustaining a dialectical critical realism that is redefined in and from the Global South involves a particular-universal reconstruction, that is, dialectic, of the potentialities of theoretical practice that until the present has demonstrated its disruptive and imploding potential of European reason: Marxism. The disruptive is observed in its privileged position in the historical social courts of the world system since the end of the 19th century, and its implosion activity manifested by its ability to reveal the closures and obturations of capitalist modernity. Only the history of the innumerable times in which Marxism has been declared dead and buried in just over a century and a half of existence speaks of its destructive vigour concerning the theoretical practices that the political economy of European morality has built and rebuilt in that period. The STS are the result of the implosion of colonial reason, they are the result of its very instruments that create the edges of its catastrophes.

It is in this context that we present the current work, at the crossroads of the theoretical spheres to which we have alluded, which are the conceptual "instruments" to address some of the central aspects of the multiple challenges to construct theory(s) from the Global South. We will refer to the tools that allow us to account for mechanisms, processes, and experiences, making their connections explicit from a dialectic that captures the epistemic, the theoretical and the critical-instrumental, leaving for another opportunity the analysis of the ontological and methodological spheres beyond the allusions we make of them.

Inscribed in the above, we will expose how there is in the STS an epistemic, theoretical, and methodological configuration that can be understood in the joints and disarticulations, in the connections and disconnections, in the proximities and distances that they provide: a logic of the *germinal explanation* of the structures; a *Moebesian, spiral, reticular,* and *dialectical constitution* of social processes-phenomena; and chromatic analogies for the understanding of social practices for the construction of substantive information from an approach that retakes sensations as a starting point to listen to multiple voices.

In the context of what has been written so far, and within the framework of the proposals for a germinal explanation of the mechanisms, Moebesian, spiral and reticular analysis of the processes and the use of a theory of colours to reconstruct the interaction experiences, a possible agenda for the development, discussion and validation of STS can be summarized as follows:

1. STS is post-independence because it implies assuming the consequences of the "new" forms of the colonial in our societies. In this context, it is a priority to analyse the social forms of subjection from the logics of independence and autonomy.
2. STS accepts that the battle of knowledge is a critical action of criticism stabilized and coagulated in academic power, and that from this follows an immanent criticism of "critical thinking" stagnant as an ideological practice of colonial postmodernity.
3. STS promotes the discussion on the unequal distribution of knowledge and the ability to narrate one's history involved in the plots constructed by the geopolitics of knowledge, planetary predation and the daily counter-insurgency operations of the international repressive apparatus.
4. STS promotes the theoretical practices necessary to go from being mere reproductive throats of other voices to the construction of plural voices capable of elaborating rebellious and insubordinate worldviews.
5. STS maintains that it is theory insofar as it is a scientific explanation of the world with an emancipatory purpose, pointing out the counter-expropriation, interstitial and "interdictional" forms that social practices assume.
6. STS encourages the inclusion in said explanation of the dialectic between the constitution of the Nero complex as a usurper, the Columbus complex as a benevolent conqueror and the Pontius Pilate complex as an indifferent colonizer, and its tensions with current forms of enjoyment, happiness, and hope.

As it is possible to note, the elaboration of southern social theories involves a postcolonial epistemology and a decolonial attitude towards conceptual logic.

In the theoretical-epistemological context described, it is important to explain two more components of the perspective maintained in this book: the approach to dialectics and the conceptualization of the world image as a substrate of the conceptual set that is elaborated through the book.

In principle, the notion of dialectics that has been used here makes us think of three possible meanings: firstly, it can be identified with a processual and relational vision of the phenomena that structure the social world; secondly, it indicates a propaedeutic process for the knowledge of the social; and finally, in a less strong way, it indicates the moments of articulation of the phenomena that occurred in the social structuration process. The first sense addresses the facet of constitution of the social world that has generally been limited, but not resolved, in the diachronic/synchronic dichotomy. That is to say, the social cannot be exhausted in the knowledge about simultaneity or in the knowledge about becoming, for this it is necessary to incorporate a look that allows us to capture what is simultaneous and what there is of change in the social world. The dialectic is thus a feature of the social that involves its moments of change and permanence crystallized in what is the process of relationality of the agents and social actions. In this direction, the dialectic of structures and praxis can be identified with the basic characteristics of the relationship between agent and structure, between agents, and between actions, which gives social reality that quality of appearing as simultaneous and changing at the same time. This feature of the social constitution leads us to wonder how the moments of production and reproduction of social reality emerge. In this way, it is possible to understand that if one looks carefully, both the structure and the praxis are both means and products of human agency. Therefore, it will be possible to appreciate how the dialectic is a cognitive and argumentative path to enter a social reality that is presented as complex and indeterminate. That is to say, dialectical here means the way in which we introduce ourselves to the knowledge of a reality that is simultaneous and changing, produced and reproduced by agents. It is to look for relationality and processuality, to give it a meaning so that it can be understood in a very close direction with the work of interpretation proposed as a hermeneutical task; dialectics is the recognition of the multiple and complex voices that emerge, but which it is possible to bring into dialogue in a procedural rationality. In this sense, dialectics is rational argumentation and is a form of knowledge and epistemic assumption of scientific knowledge of the social. But a third meaning is also suggested, which although it is "weaker" than the previous ones, is the one most directly used in the elaboration of theory in social sciences, it is the one that indicates the processes of articulation between phenomena and that cannot be identified with the characteristics of these phenomena neither with those of the agents, nor with those of the agency. It is precisely the same moment of articulation that must be conceptualized, taking into account the features of what is articulated, which provides a new reality not identifiable with those that precede it.

This book attempts to provide the necessary elements to understand, at least partially, a world image of the current situation of social structuring based on the colonization process of the inner planet. The images of the world imply the presupposition of a particular relationship between agents, time, space, and reality; giving theories a pre-comprehensive horizon. In other words, it allows social theorists to choose between different ways of assuming the way of existence of the inhabitants of the social world.

Among the elements provided by these definitions, it is possible to highlight the following: (1) the reference to the way of existence of certain entities; (2) these entities are carried by language but inhabit the world; (3) the above implies the determination of a way to understand what the world is; (4) furthermore, that this "way" of understanding the nature of the world can be subject to analysis or criticism; and (5) it is accepted that both science, or rather scientific theories and everyday life, implies an ontology.

Thus, it can be perceived how theories contain certain figurations of how the world is, and also that asking about them differs from questioning how we can know that world. In this context, by Image of the World in relation to theories in Social Sciences, it will be understood here preliminarily the set of presuppositions about the "mode of existing" of the agents, structures, time/space and their relations with social reality, which constitute the alluded theories.

To understand an IMAGE OF THE WORLD, we must recognize the following assumptions about its constitution:

a) it also implies a vision of what are the resources that agents use to relate, distinguish themselves, and hierarchize themselves, among which we can highlight: knowledge, information, power, and wealth; here introduced in Chapter 1.
b) on the other hand, it establishes which are the subjects/objects of another species with which man shares his environmental horizon and how this relationship occurs; this is also thematized in Chapter 1.
c) includes a vision of time/space, which can be expressed in the combination of some of the following pairs: change/stability, transformation/reproduction, horizon/region, and process/mechanism that is used in this book, and captured in Chapters 4 and 6.
d) the image is a notion of a person, which involves decisions about what is intuitively usually indicated as a subject, and how these subjects are produced and reproduced; these matters are discussed in Chapter 5 in terms of the "dialectic of the person".

It requires these elements of an image of the world to acquire a specific form. Definitions about "relationships," that is, regarding how they are connected, require the articulation of a "vision of the other" and a "painting of the existing" that provide a figure for such relationships and the boundaries in which they are to be interpreted as significant.

The central characteristic that differentiates these images of the world that theories carry from those that we carry as subjects is their radical provisionality and uncertainty, as discourses tied to an argumentative rationality. The images of the world of theories are always to be founded, to be rationally argued; the images that we carry as subjects are always there as a certainty of the common stock of knowledge that nests in our lifeworld.

The dialectic of the images of the world that is identified, described and analysed in this book has been organized in such a way that it is possible for the reader to grasp the theoretical/epistemic starting points (Introduction); the central features of three possible diagnoses of the contexts in which the phenomenon analysed in the book is included (Chapter 1); two of the most relevant ways of occupying the inner planet: endocrine disruptors and nanotechnology (Chapter 2); the impact on politics of the senses of colonization that have been referred to

(Chapter 3); some of the mechanisms, processes and experiences that constitute the social structures where colonization especially unfolds (Chapter 4); the presentation of a geometry and dialectic of the person as individual/actor/agent/subject/author articulation modalities (Chapter 5); the collective practices and their meanings in terms of connection with the conflictual networks that the situation described unfolds (Chapter 6); and finally the dispute over preferences and values in the midst of which the colonization of the inner planet takes place (Chapter 7).

I.3.2 *Sorority, fraternity, and humanity*

As a horizon of general understanding, this book is written knowing that as Sara Ahmed says:

> Intersectionality: let's make a point of how we come into existence. I am not a lesbian one moment and a person of colour the next and a feminist at another. I am all of these at every moment.
>
> (Ahmed, 2017: 230)

This book tries to be alert to the fact that the century in which we live is configured beyond all binarisms and that it cannot be understood without the intersections of experiences and sensibilities of class, ethnicity, age, and gender. In this context, this text pays specific attention to the articulation between sorority and fraternity as logics of sharing, as logics of equality, and as logics of happiness. Both sorority and fraternity dispute what is in the paternalistic, phallocentric, and patriarchal society of dictatorial authoritarianism and that undermines the autonomy of people. A community is by definition intersectional, it is a space of sharing between genders, post-generic, it is an amalgam of genres, configuring itself in a line where the space of division has not yet been born, in the vectors of non-division. It is a space and a path where sorority and brotherhood meet, where the co-experience, the call and the cry of sorority that leads to a radical sharing and the legacy and the dictum of the brotherhood meet and complement each other. In this context, there is no community project without articulation for equality, thought from sorority and fraternity, thought in and from the capacities and skills of each one of the members of the community, beyond their position and generic disposition, and beyond their gender options and identities. Sharing from sorority and fraternity lead to radically communal equality where no one stands out above the other, where no one is worth more than another, where there is no person without a community, and therefore equality is enshrined. Not only equality before the law, not only access to goods, but profoundly equality to build an identity project, equality to build a life project, equality in the distribution of nutrients, and equality in the configuration of the future. It is this relationship between sharing and equality in sorority and fraternity that leads to the rediscovery of humanity as the result of vectors of coexistence of all species, all genders, and all life projects.

By understanding other living beings as co-inhabitants of the planet, we imply a path towards a post-speciesist gaze in a co-living between humans and co-living with other species; that is: the position in the cosmos is not only of man, but of all living beings. In this sense, another very relevant element for this book is to inscribe the description of the colonization of the inner planet beyond any speciesist position, emphasizing the need to elaborate social sciences where all living beings are participants of the planet as inhabitants of it. We must go beyond speciesist discrimination of those who do not belong to the human species, a position that can also be characterized as anthropocentric speciesism that must be transformed into post-speciesism (Singer, 1990, 2006; Horta, 2009).

It is here where the geometry that is formed between sorority and fraternity and humanity becomes a springboard for happiness, not as enjoyment through consumption, not as self-centred use of unequally distributed resources as deemed unilaterally appropriate, but as a life project where trust, love, and reciprocity allow us to talk about the future, that is, it allows us to have hope. In this context, we are happy. After all, we are with others because we share in an equal way the possibility of the option of gender, identity, and the elaboration of plural futures. A radical community commitment is a co-living that becomes co-authorship of the world; we live with others, equally sharing the possibility of building happy worlds. The consequences of the old struggle for fraternity and the current struggles for sisterhood are updated, instantiated, and become practical in the construction of humanity where all forms of discrimination, inequality and authoritarianism are dissolved, and thus becomes possible the construction of a common life sharing projects of hope, identity options, and paths to the future.

I.4 Book content

The book contains an Introduction and seven subsequent chapters. The Introduction and Chapter 1 are longer, and are intended to systematize a set of assumptions and describe the scenarios necessary to understand the arguments and analysis offered in the other chapters.

The first chapter focuses on summarizing three convergent diagnoses of the current situation of the expansion of capitalism on a global scale: planetary depredation that implies the birth of a "new" religion of helplessness in the context of contemporary coloniality; the consolidation of societies normalized in immediate enjoyment through consumption, which involves the internationalization of emotionalization, the logic of waste, the politics of perversion and the banalization of the good; and the central characteristics of Society 4.0 and the connection with the politic of sensibilities are also made explicit.

Chapter 2 is dedicated to one of the two most important facets of the colonization of the inner planet, that is the management of the design, production and reconstruction of the material conditions of the body's existence in the central nervous system, the immune system and especially with the brain. The argumentative strategy is to first offer a synthetic presentation that makes explicit of some of the central phenomena of the colonization of the inner planet; then the power of endocrine disruptors is explained, and the strategic place of nanotechnology is shown.

Chapter 3 makes explicit how the senses are both tools and bridges that human beings use to connect with the world. Beyond their number and complexity, it is possible to establish how the five basic senses have been invaded, colonized, designed and reproduced under the double logic of political and economic coloniality in the last part of the 20th century.

Chapter 4 aims to carry out a triple task: (a) present a theoretical/epistemic look at what can be understood by structures at present, (b) characterize the appearance of the "subsiadano" as an expression of the modifications of the position of citizen, and (c) to make explicit the process of internationalization of the emotionalization that plays the role of a background to the current social structuration processes on a planetary scale. The chapter follows a spiral logic where the theoretical, the analytical, and the descriptive are connected to show at least some of the salient features of the context of the colonization of the inner planet; introducing, partially, the appearance of new forms of dependency as the predominant colour of emotionalization.

Intimately linked to the central objective of the book is the view we take of the colonization of the inner planet, and as we have already anticipated, in Chapter 5 we make a first approach to dialectical personalism as an identity structure to understand the meaning of the human being today. In an intersectional, "non-speciesist," and decolonial vector spectrum, a review of the concepts that sociology has used and reuses to designate people is carried out: individual, subject, actor, agent, and author.

Chapter 6 shows how collective actions are connected with the politics of the senses and the structural features of the current process of colonization of the inner planet. To achieve this objective, the following is carried out: (a) the configuration of emotional ecology is presented; (b) the connection between social protest and conflictive networks is sketched; (c) expressive and "aesthetic-in-the-streets" resources are described; and (d) a synthetic conceptual approach about interstitial practices, collective interdictions and experiences of affirmation is provided, and finally messages, symptoms, and absences are shown.

The last chapter maintains that to elaborate a social theory that decolonizes the forms of depredation of the inner planet, it is necessary to thematize the structure of the political economy of morals, understanding that it is in and through it that such depredation is justified/naturalized in connection with the material conditions of existing. The chapter ends by discussing the need to emphasize research and criticism of what is understood by virtues within the framework of a society normalized in immediate enjoyment through consumption. It seeks to emphasize the need that every social theory has to make critical the political character of ethics, aesthetics, and morals.

I.5 Some early conclusions

This last section tries to summarize, in a provisional way, some of the axes that order the theoretical look that we intend to offer on the colonization of the inner planet and its consequences.

The book maintains that it is possible to identify three poles of the colonization of the inner planet that in different phases of progress are reaching the shores of the new world carried by the new ships. The first pole is linked to production and distribution in manufacturing, consumption, and the enjoyment of endocrine disruptors used in hundreds of products and processes. The second pole is the design and genetic manipulation through the use of DNA that is easily identifiable today, encompassing both animal, plant, and human genetic maps and management. The third pole is the use of nanotechnology as a tool for exploration, intervention and transformation of the bodies/emotions, constituting a new form of colonization and commercialization.

The inner planet is colonized in three ways: (a) there are vehicles that navigate it, (b) there are assets that are absorbed/preyed on, and (c) there is communication with the metropolis it occupies. Cells, molecules, and atoms are explored, occupied, and appropriated by the "external", social, subjective, and natural worlds that transmit their metabolic fracture[3] to them.

The politics of the senses is the expression of the colonization of the inner planet in and through the senses that operates through three concurrent pathways: (a) by artificial mechanisms, (b) by writing the biological structure, and (c) by the interface between a and b. The politics of the senses play a triple role in contemporary societies: (a) it impacts on the geometry of bodies and the grammar of actions, (b) it is a nodal component of the rules and norms that shape the institutions of the state, the market and civil society, and (c) are part of the central "architecture" of what in Chapter 5 is called the dialectic of the person.

In this vein, a question emerges: are the transformations in our scopic regime transformations in our value system? If all aesthetic transformations correspond to certain ethics and, in turn, certain politics, then the answer may be that they are. Accepting that we are "feeling-thinking" beings (in the sense of Fals Borda) leads us to wonder about our "video-touching" condition as producers of sensibilities that enable us to know/sense the world. The challenge for the social sciences of societies standardized in immediate enjoyment through consumption in the context of the revolution 4.0 still is how to connect/disconnect science and politics.

In this book, it is possible to see how "the political" is intertwined with "the emotional." The globalization of emotionalization serves as the central axis of the current metamorphosis of relations between state and capitalism, between politics and market, and between "ideology" and marketing. The colonization of the inner planet is part of a Moebesian strip for which a special form of citizen appeared in the 21st century, attesting to the close connection between consumption, state subsidy, and citizenship.

The book makes evident the emergence of a "new" position of person and citizen in the context of the society normalized in the immediate enjoyment through consumption: the assisted citizen. One of the main consequences of the connections between social policies and the politics of sensibilities is the "creation" of a modality of subjectivity/personality based on the close relations between consumption, assistance, and enjoyment.

In this context clearly appear an internationalization of the emotionalization regime that involves the globalization, local massification, and global spectacularization of the commodification of the elaboration, management, distribution, and reproduction of emotions, something that has become the competitive and motivational core of capitalism. Feeling, experiencing, having an experience, and connecting with objects/subjects are the most demanded and produced goods in current capitalism. The difference of this stage/moment of the dialectical structuration of capitalist restructuring is that these emotions are not qualities of an object; on the contrary, they are the objects that are requested, acquired, consumed, and discarded.

In the framework developed here, the book finds in collective actions a set of practices that allow us to perceive the actions and reactions to the processes of planetary deposition and predation of the inner planet: topologies of rejection. The topologies of rejection are forms that make up contradictory fields of forces, morphologies of denial, and a Moebesian tape of denials. Saying "no", maintaining distance, and denying resignation, are practices that shape life lived in autonomy and are perhaps the key to the future marches of the collective. Neither interstitial practices, nor collective interdictions, nor affirmation experiences alone, are sufficient for an inaugural act of autonomy. We will have to strain the subtlety of the observation to capture the new situations where topologies of rejection are developed from the Moebesian tension between the three.

In the colonization of the inner planet where the transfer of the metabolic fracture to the same body/emotion is taking place as the "last" territory of conquest, the dispute and the reconstruction of those designated as individual/actor/agent/subject/author is one of the axes of the social structuration processes. Individual/actor/agent/subject/author are at the same time modalities of designating the dispositions of the person, and therefore its geometry and its contingent tensions of existence are captured/understood from its dialectic. While the colonization of the inner planet is taking place, understanding these facets of the bodies/emotions as their "cobordant" instances (*sensu* Thom) allow us to understand their limits and analytical possibilities.

This geometry of the person is stressed and shaped as a mode of existence and criticises the political economy of morality; it involves an alternative sensibility that challenges the acceptable and accepted politics of sensibilities for its claim of a single and closed totality. To exist, to live, and to sensitize are the three moments that emerge as a consequence of the dialectic between the appropriation of the phylogenetic and ontogenetic that the individual carries, the criticism of the sociabilities and experiences that the actor must interpret, the empowerment of the agent's skill in redoing the action, the autonomous reflexivity of the subject who sees himself as the axis of his identity, and of the collective and of the author as creator of new possibilities of making life.

Both the geometry of the person and their dialectics leave us at the doors of a final discussion: what are the priorities/values and how can they be understood in societies where the colonization of the inner planet is increasing day by day?

The game of the dialectic of the person, wherein the possibility of denying the claim of totality the political economy of morality appears, this possibility implies a reflection on the central axes of some interactions created from listening, dialogue, and hope as its opposite.

As will be argued in the last chapter, in our days, the colonization of the inner planet is being systematically completed and in the face of this, a radical critique of the available knowledge is needed to scrutinize the procedures, paths, and vehicles of colonization. It is within this framework that the development of social sciences that can perform this task from a trans-disciplinary perspective becomes imperative.

After these notes, this writing sketched as the record of a voyage, making evident the features of a still incomplete log, there is still much to do/say/think. The social sciences, in commitment to utopian thought, have a long way to go. Weaving hope with the possibilities of caring for ourselves as close beings who co-inhabit a time/space is today one of the central challenges of these social sciences. Taking back the virtues and rebuilding utopias away from prejudices and encouraged by the daily force of interstitial practices is a central task for the university and the scientific system of the planet.

The task requires some social sciences commitments, involved with the criticism of some rationalities centred on the bureaucratization of knowing/doing/feeling and registering in the dispute for the recovery of the potentialities of virtues, building alternative models for knowing (us). In this direction, an alternative model of knowledge is above all a collective task, where knowledge is a lifestyle that contributes to the democratization of the colonization of the "natural," social and internal worlds. It is a knowledge that bets on the potentialities of a radical democracy of intimacy , autonomy in life decisions, and human emancipation. The book closes with a call to hope as a radical critical instrument in response to the colonization of the inner planet.

Notes

1 This connects directly to what unfolds in Chapters 2 and 3 in terms of the colonization of the inner planet.
2 See Chapter 4.
3 For the concept of metabolic fracture CFR Machado Aráoz, Horacio (2015) Machado Aráoz, Horacio y Rossi, Leonardo (2018) Machado Aráoz, Horacio (2018).

References

Ahmed, S. (2017) *Living a Feminist Life*. London: Duke University Press.

Anzaldua, G. (2012) *Borderlands/La Frontera: The New Mestiza*. San Francisco: Aunt Lute Books.

Castro-Gómez, S. and Mendieta, E. (Eds.) (1998) *Teorías sin disciplina, latinoamericanismo, poscolonialidad y globalización en debate*. Mexico City: Miguel Angel Porrúa/ University of San Francisco.

Coronil, F. (1996) "Beyond Occidentalism: Toward Nonimperial Geohistorical Categories", *Cultural Anthropology*, 11, (1), pp. 51–87.

De Sousa Santos, B. (2018) *The End of the Cognitive Empire: The Coming of Age of Epistemologies of the South*. Durham: Duke University Press.

Dussel, E. (1993) "Eurocentrism and Modernity." In: J. Beverly and J. Oviedo (Eds.) *The Postmodernism Debate in Latin America*. Durham: Duke University Press, pp. 65–76.

Dussel, E. (1996) *The Underside of Modernity*. Atlantic Highlands, NJ: Humanities Press.

Dussel, E. (2000) "Europe, Modernity, and Eurocentrism", *Nepantla*, 1, (3), pp. 465–478.

Escobar, A. (2007) "Worlds and Knowledges Otherwise", *Cultural Studies*, 21, (2–3), pp. 179–210. https://doi.org/10.1080/09502380601162506.

Guha, R. (1963) *A Rule of Property for Bengal: An Essay on the Idea of Permanent Settlement*. Paris: Mouton & Co.

Guha, R. and Spivak, G. C. (2002) *Subaltern Studies Modernità e (Post)colonialism*. Verona: Ombre Corte.

Horta, O. (2009) "Ética Animal: El cuestionamiento del antropocentrismo: Distintos enfoques normativos", *Revista de Bioética y Derecho Publicación Cuatrimestral del Master en Bioética y Derecho*, 16, abril. Avaiable at: http://www.bioeticayderecho.ub.es

Lander, E. (Ed.) (2000) *La colonialidad del saber: eurocentrismo y ciencias sociales*. Buenos Aires: CLACSO.

Lander, E. (2002) "Los derechos de propiedad intelectual en la geopolítica del saber de la sociedad global." In: C. Walsh; F. Schiwy and S. Castro-Gómez (eds) *Interdisciplinar las ciencias sociales*. Quito: Universidad Andina/Abya Yala, pp. 73–102.

Lugones, M. (2002) "Impure Communities." In: P. Anderson (Ed.) *Diversity and Community: An Interdisciplinary Reader*. Oxford: Blackwell, pp. 58–64.

Lugones, M. (2003) *Pilgrimages/Peregrinajes: Theorizing Coalition Against Multiple Oppressions*. Lanham: Rowman & Littlefield.

Machado Aráoz, H. (2015) "Marx, (los) Marxismo(s) y la Ecología: Notas Para un Alegato Ecosocialista", *Revista GEOgraphia*, 17, (34), pp. 09–38. http://www.uff.br/geographia/ojs/index.php/geographia/article/view/837.

Machado Aráoz, H. (2018) "La insustentabilidad del capital. Ecología política del Sur, crisis ecológica/civilizatoria y la cuestión de las alternativas." In: M.L. Eschenhagen and C.E. Maldonado (Eds.) *Epistemologías del Sur para germinar alternativas al desarrollo*. Bogotá: Universidad del Rosario–Universidad Pontifica Bolivariana.

Machado Aráoz, H. and Rossi, L. (2018) "Extractivismo Minero y Fractura Sociometabólica. El caso de Minera Alumbrera Ltd., a veinte años de explotación", *RevIISE*, 10, (10), octubre 2017 - marzo 2018. Argentina. pp. 273–286. ISSN: 2250-5555. Available at: www.reviise.unsj.edu.ar

Melucci, A. (1996) *The Playing Self: Person and Meaning in the Planetary Society*. Cambridge: Cambridge University Press.

Mignolo, W. (1995) *The Darker Side of the Renaissance: Literacy, Territoriality and Colonization*. Ann Arbor: University of Michigan Press.

Mignolo, W.D. (2007) "Introduction", *Cultural Studies*, 21, (2–3), pp. 155–167. https://doi.org/10.1080/0950238060116249.

Quijano, A. (1988) *Modernidad, Identidad y Utopía en América Latina*. Lima: Sociedad y Política Ediciones.

Quijano, A. (1993) "Modernity, Identity, and Utopia in Latin America." In: J. Beverly and L. Oviedo (Eds.) *The Postmodernism Debate in Latin America*. Durham: Duke University Press, pp. 140–155.

Quijano, A. (2000) "Coloniality of Power, Eurocentrism, and Latin America", *Nepantla: Views from South*, 1, (3), pp. 533–580.

Quijano, A. and Wallerstein, I. (1992) "Americanity as a Concept, or the Americas in the Modern World-System", *International Social Science Journal*, 134, pp. 459–559.

Singer, P. (1990) *Animal Liberation: A New Ethic for Our Treatment of Animals*. New York: Random House.

Singer, P. (2006) *In Defense of Animals: The Second Wave*. Malden, MA: Blackwell Pub.

Tyagi, R. (2014) "Understanding Postcolonial Feminism in Relation with Postcolonial and Feminist Theories", *International Journal of Language and Linguistics*, 1, (2), pp. 2374–8869.

1 Diagnosis of the 21st century

1.1 Introduction

Every social theory is dependent on the social context of its production, and therefore, at least in a synthetic way, it is necessary to clarify in which scenario emerged the need to analyse the colonization of the inner planet as a theoretical starting point that approximates a better understanding of the 21st century. It is within that framework that the first chapter focuses on summarizing the said starting point. In other places (Scribano 2015, 2017, 2018, 2019; Scribano & Lisdero, 2019), we have examined how the social structuration processes on a planetary scale implies, on the one hand, the consolidation of societies normalized in the act of immediate enjoyment through consumption. Such normalization involves the internationalization of emotionalization, of the logic of waste (LoW), the politics of perversion (PoP), and the banalization of good (BoG). On the other hand, we have exposed the birth of a "new" religion of helplessness in the context of contemporary coloniality, and how it connects with the depredation of all types of energy from the population of the planet. These two diagnoses must be complemented with the synthesis of the basic features of Society 4.0 that, as will be clearly seen in Chapters 2 and 3, plays a fundamental role in the colonization of the inner planet.

This chapter aims to provide an approach to the diagnosis of the current situation of capitalism on a planetary scale that results from the articulation of three elements: (a) a sketch of the configuration of a global predation device; (b) a summary of the central features of the consecration of societies normalized in enjoyment through consumption; and (c) the schematic presentation of the components of the Society 4.0.

The sketch of the situation of predation, dispossession, and expropriation on a planetary scale of all types of energy from the earth (biological, geothermal, mineral), water, air, light, and especially from bodily energy, seeks to develop a scenario of dispossession where the consequences of the colonization of the inner planet can be inscribed.

This summary of society normalized in immediate enjoyment through consumption seeks to paint a picture of the world of the emotionalization process that at a global level socializes new characteristics that qualify accepted and acceptable sensibilities.

Finally, a schematic presentation is made of how the consolidation of Society 4.0 converges with what has been previously described as the inscription surface of the set of practices that will be described in Chapters 2 and 3.

1.2 The colonial situation in the 21st century

Starting from the sociology of bodies/emotions that was synthesized in the previous chapter, and following the results of numerous empirical investigations, the following interpretive nodes of today's social structuration processes have been reached:

a) The current situation of the Global South in particular, and of the planet in general, is characterized by (1) a great process of depredation of common goods especially that of the corporal/emotional energy, (2) the consolidation of social bearability mechanisms and devices for the regulation of sensations on a planetary scale, and (3) a piece of immense repressive machinery that has been elaborated on the basis of fear and violence. In the context of the situation described, a "new neocolonial religion" has been developed whose dogmas of faith are mimetic consumption, resignation, and diminished humanism. The above-mentioned dogmas connect with the massification of a "sociodicy of frustration" that obliterates disruptive practices and reproduces the consolidation of the present political economy of morality.

b) Within the framework above, it is possible to note the consolidation of societies normalized in immediate enjoyment through consumption that implies the consecration of a sacrificial and spectacularized structure of a cultural economy for global emotionalization.

c) In respect of the breakdown of the neocolonial religion, there is a set of interstitial practices (love, reciprocity, joy) that deny the truth regime of the political economy of morality and which, in turn, are inscribed in the context of a set of disruptive practices such as collective "interdictions" and topologies of rejections.

Capitalism, as a producer of a politics of sensibilities whose function is to provide a cement that reties the effects of individualization/fragmentation, the spread of instrumental rationality as a logic of social interaction being the result of the disenchantment of religious ethics, and the constitution of a political economy of morality based on shared beliefs about the rules of conduct cropped in the shape of the elementary principles of the market as an explanatory mechanism for the behaviour of human being, makes it possible to understand sociability as a religion. This is how it may be understood that capitalism as religion provides basic images of cohesion, promotes a form of joint action, and structures a set of beliefs about the origin, development, and purpose of both cohesion and action.

It is in this sense that the structures of the political economy of morality are now a secular religion of all subjects who live within the capitalist system, and especially of those expelled homeless people in the Global South. This political economy of morality is the consequence of the consecration a "new" politics of sensibilities.

Facing the processes of expulsion and surplus expropriation, merchandise worshipping-turned-fetish operates as moments that reunite, rebind, and reconstruct what in these processes is an experience of disconnection and loss. Moreover, these moments are elaborated on the grounds of the expansion of capital on a global scale whose results involve just its investment: emptiness, loss, and decay. To understand the contradictions that nest on these results and how they relate to capitalism as religion, it is necessary to expose, at least schematically, the central features of the aforementioned expansion of capital. There are three axes of the same Moebius tape that cross, helically and dialectically, the current state of capitalism at a global and regional level: the predatory practices of the commons, the development of social bearability mechanisms and devices for the regulation of sensations, and the redefinition of the repression-militarization of societies.

First, in its various phases, imperial capital has always had as its objective to ensure the long-term conditions of its reproduction system-wide. At present, the monopolist concentration of capital becomes a device of air extraction of the present to manage the air of the future. The living force of capital, human beings turned-mere "bodies-in-work" for the enjoyment of the few under the fantasy of a desire of all, needs to ensure the highest rate of ecological ownership in order to maintain in the medium-term the (changing) structure of the ruling classes. In this regard, the location, management, and treatment of water sources worldwide are among the edges of the predatory extraction strengthening its metamorphosis under conditions of inequality. No water, no bodies, or food reproduction systems; the biogenetic safeguards that are the necessary and sufficient conditions of the appropriation of the future. The consolidation of air and water extraction (in the context of processing, storage, and distribution on a global scale) is based on the need for ownership of land that comprise and produce these two core elements of life. Jungles, forests, and fields must be secured by the alliances of the fractions of the national ruling classes through guarantees of national states to private ownership, that is, privatized and globalized international corporate environmental management. Among the many objects/materials of predation are the "rare earths" that, as we will see in Chapter 2, constitute one of the central axes of the "transmission" of the metabolic fracture to the colonization of the inner planet, and many of them are keys to the information storage industry.

In the same direction, the other edge of the extractive machinery is energy in all its forms, from oil to bodily energy, made socially consumable and available. Beyond the fatal process of extinction of these basic energies for capital, its regulation is currently the centre of reproduction in the short term. Therefore, a critique of the eco-political economy is an important and indispensable step for understanding imperialist expansion. A constituent element of such a review is to make visible how politics of bodily energies meet, are revealed and are written. The tribulations that numb bodies through social pain are among the primary means for an unequal appropriation of the aforementioned bodily energies.

Second, for the current phase of imperialism, the production and handling of the regulation of expectations and the avoidance of social conflict are essential. Such management is ensured by the social bearability mechanisms and sensation control devices, to which we shall return later, these being the key components of neocolonial religion.

Third, imperial expansion involves in various ways the militarization of the planet. There cannot be a balanced operation of extraction equipment and devices for the regulation of sensations without a repressive apparatus, a disciplinary and global control that transcends mere military occupation. This global repression aims to uphold the rule of neocolonial surveillance, given the paradoxical reorganization of the compositions, class positions, and conditions in complex spacetimes with centrifugal (which move away from the centre) and centripetal (towards the centre) movements of the various ways to resist the energy expropriation and regulation of feelings. In addition, the potential militarization of all conflicts in geopolitically dependent systems is due to the metamorphosis of concentrated financial capital, the redefinition of corporative "patterns of accumulation," and both the fragmentation and unity of the expropriation.

In this it may be understood, at least partially, how imperial expansion, characterized as an extractive apparatus of air, water, land, and energy and as a repressive military machine, is sustained and reproduced by, among other factors, the production and handling of the regulation of sensations and mechanisms of social bearability.

In the same way, it can be understood how the linking texture between particularizations, individuation, and locations are connected with the inequalities in the current state of the expansion of capital and imperialism, dependence, and the colonial situation. As we have argued elsewhere (Scribano 2010):

a)   When there are social groups that centralize the enforcement capability focused on the definition of needs, desires, and actions constituting a political economy of morality which consecrates the surplus expropriation, thereby avoiding any form of autonomous practices, it is a form of imperialism.
b)   When there is a pattern in relations between territories, nations, and states that socializes the destructive effects of the accumulation processes of environmental assets, conditioned by the status of high productivity fields, structured through global ruling classes' connections, we face a situation of dependency.
c)   When there is class segregation behind "walls" that contain and reproduce the moments of expropriation and dispossession consecrated by racialization, there is a colonial experience.

In the aforementioned context we can understand "colonization" as an act that starts and reproduces in the (colonial) cities with the following characteristics:

To colonize is to occupy. From this perspective, the city shows how the current phase of capitalism reconfigures its power to and from the expulsive and segregationist urban maps.

To colonize is to expropriate. Class occupations of cities operate as a cumulative de-possession of the capacities for living.

To colonize is to inhabit another's space-time. A life lived "from the face" of closeness and a distance that paralyses out of fear is the mark of the edges of a city and its mental walls.

To colonize is to have the power to decide on the lives of others. The neocolonial city is the iterative realization of frames of the imposition of wills over others in terms of total heteronomy.

It is in this context of colonial subjection that alterations are part of the differences and hierarchies between individuals and social groups in the Global South. The elementary forms of class modulations are pluralized and concentrated; they multiply and cluster around the dialectics of the colonial, rebuilt as a scheme of every social formation.

In the context of the production and reproduction of the preceding colonial experience, a way of supporting and "participating" in it emerges which contains all the features of religion as we have outlined above.

The various facets of the expansion process of capitalism on a global scale have required different ways of expressing the political economy of morality. From bourgeois waste through to ascetic thrift, leading to the development of racial identity as a parameter of behaviour, are some examples of the aforementioned transformations.

Part of the cunning of colonizing reason was to declare dead every religion, leaving alive only the modern and modernizing religion of goods. Among the first worship practices was the belief placed in the accumulation of wealth as the only path to a decent life, and this was accompanied by the accumulation of power and knowledge as social forms that guaranteed a full life. In this context, capitalist modernity inaugurated the curious religion that does not retie but individualizes fragments and "self-refers" social relationships and establishes a set of "beliefs" that had to be deconsecrated in the reproduction of the secular bourgeois state.

In the empirical and theoretical research, we have undertaken previously, we have developed a characterization of what we call the trinity of neocolonial religion (Scribano 2009a, 2009b), as a reading of the current state of incorporation of a political economy of morality. This is a religion that has its tables of the Law, its liturgies and pastoral, in line with the social structures involved in the current phase of imperial expansion.

A sociology of indeterminate subjection involves an acceptance that if we claim to know the existing patterns of domination in a society, we must consider: what are the distances that society imposes on its own bodies? How they are consequently branded? And how are their social energies made available? Thus, the politics of the bodies, that is, the strategies that society agrees to apply in order to answer to the social availability of individuals is a chapter, and a not unimportant one, in the structuration of power. These strategies are tied to and are "strengthened" by a politics of the emotions aimed at regulating the construction of social sensibilities.

It is precisely in the weft of bodies, and the embedding of the politics of emotions in the colonial experience, that there appear and are displayed the features of neocolonial religion. A Trinitarian constitution of the religion of helplessness[1] is elaborated in this context.

Currently, this can be seen in the emergence of a neo-religion of helplessness. This (institutional) policy ought to create the new religion of neocolonial dependent countries to replace the old trinity of "industrial religion" based on unlimited production, absolute freedom, and unrestricted happiness, with the trinity of the expelled composed of mimetic consumption, diminished humanism, and resignation. It is a religion whose liturgy is the construction of social fantasy, where dreams play a central role as the kingdom of heaven on earth, and the frustration of sociodicy's role of narrating and making present and acceptable the phantasmic hells of the past felt like a present continuous.

1. *Mimetic consumption* implies a set of practices, transversely connected, that are developed in the form of absorption of properties of an object as an act of appropriation shaped by fantasizing on the qualities of the subjective carrier of such action.
2. *Diminished Humanism* is a relationship of the suture of absences registered in a subject carried out by one or more other subjects that leave intact the processes that cause such absences.
3. *Resignation* is a way to make bodily (like embodiment) the non-future (un-future) of expectations and desires of a subject, constructed as a result of the acceptance of the limits imposed by their material conditions of existence as a closed totality.

Social mandates are installed as the "new tables" of the Law: "Consume to be happy," "Be good at some time in the day," "Resign yourself! Because that's all you can do…" are some of those mandates. From – and forth – this Moebius trinity between consumption, which makes us be someone; between disminished humanism, that the only one who benefits is the giver; and between resignation, that all it does is to seek acceptance of the limitation of the capacity of action; there are social consequences of collective multiplication which are ritualized and intertwined. That is to say, what is the pastoral of the religion of dependent capitalism? There are two elements: social synaesthesia and social ataxia. In the first, day-to-day living is enrolled in a world of hypersensibility where sensations overlap and become undifferentiated; from the second – being the inability to coordinate joint movements – subjects accede to social atomism.

Mimetic consumption has two networked poles: mimetic identification with the object and the heteronomy of enjoyment. The "enjoyment-from-the-other" is one of the most important articles of faith of the secular trinity of colonial religion based on diminished humanism, resignation, and mimetic consumption. The libidinal structure of capital is "armed" (in all its denotations) on the tripod of objectification of the other, the other as a means, and privatization of the passions. The insidious aim of this political economy of morality is for individuals who make themselves into objects, generate acceptance of oneself as ready for consumption by others, and commit to the processes of constitution of objects as "measures" of the subjects. To objectify is to put living beings (especially humans) into a state of reification to develop their status as objects that involve the objectification of life. Along the same lines, the other appears as a means to an end: the "other" is a means for "my" enjoyment. The qualities of the object become "just" a thing that can be bought, sold, accumulated, and transferred to the one that has the power to buy, sell, and build anything. Enjoying benefits of the other as object constitutes a possible path for a "me," unanchored to any other type of relationship than the commercial one. It is in this context that the *sine qua non*-condition for capitalism makes its appearance: privatize the passions to make the world of human being into a world of things. The objectification and mediation are successful only if you pay the price of a joy that leaves behind the passions, leading to multiple nets of mediation of enjoyment. The good colonist – reversing the previous religious practices – must win himself to get lost in a sea of sensations that are simply objects arranged for enjoyment, fleeting, and irreversibly instantaneous: the colonized is the first object of such practices.

Diminished humanism is constituted by a set of practices that operate as a mechanism of the suture of the differences and inequalities between classes. Such practices are characterized among other traits by reversing the place of the collective and individual, erasing their differences; diluting regimes of collective responsibility, these being replaced by the fictionalization of guilt; leaving subjects to receive donations in an iterative recipient situation; and replace the State's presence by private action and reopening private philanthropy as charitable care mechanisms. Solidarity naturalized as the potency of those who have more becomes diminished humanism as excess, resulting in the cancellation of its own foundations. As it is an outcome of social structure (as a collective), charity, is consecrated as a result of individual practice, individually sewn by another action of giving. Diminished humanism needs the acceptance of their state of need by the subjects and requires the fictionalization of social guilt without taking any collective responsibility for it. It is in this context that the neocolonial religion involves a process of redefining the content and scope of philanthropy as a social mechanism that by privatizing inequality "erases" the need for state intervention.

Resignation is a form of naturalized acceptance of the inability to rupture the "World of No" made body/flesh. In the same way, resignation involves the assent of non-truth like the social expectation adequacy regime.

The daily life of those living in the world of "No" is qualified by direct contact with the untruth of the fantasies of capital. No job, no education, and no access to health are connected directly with the "maxims" of the system such as: "whoever wants to, can work," "through striving one gets what one wants." In the context of dependent and colonial capitalism, untruth appears as a centre of gravity of fetishized subjectivity. The consecration and the naturalization of falsehood emerge as a social condition of possibility for the merchandising of lived experience.

When subjects experience fault/absences as a "normal" bond in daily life, heaven promised by the market is solved as mercantilization of the body/emotions. Life is structured in this way around that non-truth as the reverse of a systematic lie: no one owns anyone. When citizens of the world of "No" live, they do so through and thanks to what they lack being able to see what they are. Given this situation, whatever is intended as a naturalized vital goal becomes elusive and distant.

That untruth begins, through repetition, to become a centre of gravity for subjectivity through which the various figures of the phantom are catalysed. The non-truth is staged as the necessary reverse of domination, put in a state of a narrative that shows the configuration of accumulation and the "no's;" it's shown pornographically ("by-non-graphic"). The configuration of domination is made evident as such that, when seen, it is impossible to be lived in ways other than as resignation to a denied identity of the plurality of possible subjectivities. The subjects of domination are thus seen just in this way and in non-other, with everything they have: their expropriated and superfluous bodies. In this sense, the world of "No" voids the future in a being always in the same way, a continuous process of coagulation of action.

Just like in a Trinitarian unity, the three vertexes of the alluded religion interweave twisting and occupying their places in various ways. From this perspective, the logic of resignation is the starting point of blending in with the other that is attacking us, yet delights us from the colonial power. Considering another band of the complex dialectized web, consumption as identity brand (what makes us be someone), is connected with iterative diminished humanism as an iteration of lacking (the only one who benefits is the giver), and resignation as clotting/coagulating the action. Nevertheless, this neocolonial religion by definition is a torn totality, and in its attempts to present itself as something that does not have an external unity to deny it, the folds, cracks, and faults appear in a set of interstitial practice.

As stated already, in the "everyday life" of the millions of individuals expelled and discarded in the Global South there are unseen folding, interstitial and occluded. Thus, practices in daily lives are installed as the potency of the exceeding energy of predation. Some of the practices referred to are happiness, hope, and enjoyment that somehow emerge as counterparts to the axes of neocolonial religion set forth above.

Interstitial Practices are those social relations that appropriate open and indeterminate spaces of the capitalist structure, thereby generating a "behavioural" axis which lies transversely to the central vector of bodies and politics of emotions. Therefore, they are not orthodox practices, nor are they either paradoxical or heterodox in the conceptual sense given to them by Pierre Bourdieu.

Among the many ways of conceptually understanding what these aforementioned practices mean, we identify three of them here: as folds, as breaks, and as "unexpected" parts of a puzzle.

Interstitial practices nest in the unseen folds of naturalized and naturalizing surfaces of bodies and politics of emotions involved in neocolonial religion. They are disruptions in the context of regulation. They are emergencies that (rebel and) are revealed against the inertial vacuum to which mimetic consumption sets limits, the labelling of the impossibility to which resignation condemns and the enclosure implied in diminished humanism.

The practices to which we refer are updated and instantiated in the interstices, being the structural breaks through which the absences of a particular social system are visualized. These breaks are irregular spaces where individuals construct a set of relationships designed to weld the conflictual structure, but with different and multiple paths. They are welds that cross the body and the emotions promoting re-enchantment, uniting with reciprocity where there was mimetic consumption, combining the use of festive spending (unproductive expenditure) where there was diminished humanism, and expanding hope where resignation was the previous way. Interstitial Practices are "unexpected" parties that are linked, but are not part of, the puzzle that combines mimetic consumption, diminished humanism and resignation.

The pictorial metaphor of the puzzle does not represent but serves to "trigger" into looking how, in the context of specific social relations, there are others that are the "ins" and "outs" of what the figure insinuates. The interstitial then enters and leaves the puzzle in a contingent and indeterminate way as it depends on your particular structural historical configuration.

A colonial critique of the trinity means producing observability conditions about the aforementioned interstitial practices, and implies the following dialectical route: (a) moving from mimetic consumption to the observation of reciprocal exchange (beyond capital) and the gift, (b) passing from diminished humanism to the observation of festive spending, and (c) moving from resignation towards the observation of reliability and credibility (such as a systematic critique of ideology and re-semantization of hope). It is in this context that it is possible to better understand how a normalized society emerges in enjoyment as the structuring of a new politics of sensibilities.

1.3　The global normalization

The axis of economic policies of many states of the Global South is their neo-Keynesian character because the incentives and management of the expansion of consumption turn into their main tools. Credits for consumption, subsidies for consumption, and "official" incentives for consumption cross and overlap with the consolidated and continuous state of the development of capitalism in its contradictory depredation/consumption. Therefore, societies produce/reproduce themselves structured around a group of sensibilities that have as a context of elaboration the continuous efforts to "keep consuming."

As has been discussed in the later years of the past century, the new forms of intimacy, the diverse forms of modernity, the consequences of globalization, and the research around politics of emotions in the context of the Global South, all suggest the challenge of rethinking one of the main concepts of the 1960s and 1970s during that century: the normalized society. If we add to such a challenge the context of depredation of common goods, the high levels of poverty and indigence, food deficits, and strong processes of segregation and racial discrimination, the question about the volumes of existing happiness and optimism emphasize even more the urgency of thinking over these matters.

1.3.1 Normalization

In sociology, there have been different approaches for portraying the societies that emerged during the planetary expansion of capitalism. These include Weber's point of view about the connection between disenchantment, rationalization, and bureaucratization as an interpretative knot of social structuration; the analysis of the importance of instrumental rationality as the key interpretation in the development of mass societies undertaken by Horkheimer and Adorno; or the interpretation of the processes of the system in the colonization of the lifeworld sustained by Habermas. Beyond their different contexts of production these approaches – and many others – have in common the attempt to explain how and why the process of structuration of societies tends towards what we here call normalization.

Normalization can be understood as the stabilizing, compulsive repetition, nomological adequacy, and contextual disconnection of the social relations that the practices of individuals gain in a particular space/time:

a) Stabilization implies a set of processes of obstruction of modifications, avoidance of conflict, and balance of flows meant to pass through life with no impediments. Everyday life in the sociability time/spaces occurs in the middle of the multiple forms in which societies organize themselves for reproduction in the dialectics of production/consumption. It is "between" those time/spaces that colonial societies elaborate strategies of living that block/annul/reabsorb the modifications that the work of reproduction demands. One of the central points of the elaboration of the processes named above is associated with the creation of states of conflictual avoidance that lessen the agonistic power of the successive and generalized expropriations. At the same time, these two features are accompanied by structures and a politics sensibility that order the flows of experience with the intention that "life how it is" is "given for a fact." Stabilizing does not mean life doesn't change. It implies that experience instantiates in a predictable and manageable way. Expectations and managements, whose base has fluidity, occurrence, and indetermination give them enough plasticity and flexibility to produce stability.

b) Compulsive repetition is structured around iterative forms detached from self-reflexive processes, a decrease of self-government, and loss of individual and collective autonomy. The expansion of capitalism at a global scale generates different forms of absences/dependencies/addictions that have the common logic of needing to suture/fill/satisfy those forms in and through consumption. Social bearability, mechanisms, and devices for the regulation sensation are aimed to generate repeated and serial forms of satisfactions, through processes that are located in the pre-reflexive moments of action. The means and objects of satisfaction become inadvertently desired external solutions with which subjects can do "little or nothing." In this direction, the compulsive character diverts the capacity of commanding processes and "objects" towards the same "things" that in their autonomy detach the agent-capacity of the subject that is the object of its "help."

c) Nomological adequacy involves cognitive-affective processes of adaptation to guidelines that are performed in the pre-reflexive frame of action. Given the context of "autonomy loss," both stabilization and compulsive repetition develop in the setting of a permanent enlargement/adaptation of the rules of interaction – that turn normalization into an accepted and acceptable state – whose contents and modifications are not the object of public dispute, but instead are made effective in social histories made body. The political economy of morality combines a series of practices that become ways of understanding/feeling the world.

d) The disconnection of the context of social relationships is a mechanism for eliminating possible frictions in the processes of action coordination that enhance the flux of interactions. Since the so-called individuation processes, passing by the diagnostics about the loss/rupture of social bonds, up to the interpretations around fragmented societies, social sciences in the 20th century have described and interpreted the growing accentuation of unlinked present social practices. The normalization of the 21st century produces/ reproduces a "separation" between the actions of individuals achieving the modification of the notion of interaction itself, but it also separates the weavings between the actions of the same individual, among practices performed by the same individual. There is an inversely proportional relationship between a life lived "for-the-public" (Facebook, Twitter, the minute-to-minute opinions on TV, etc.), a life lived for "the-eye-of-another" and the unattachment between diverse "positions/conditions" that refer to the individual that performs them. The more it is shown, the idea of the tension between the different positions of the subject further disappears, as a synthesis that organizes everyday life. Neither the figure of the consumer, nor of the citizen, nor of the producer, nor the spectator (in all their versions of gender, age, ethnicity, etc.) is connected to each other. What is done at home, in the street, at the park, in the market, etc., is separated in such a way that their incompatibilities, contradictions, and paradoxes are minimized, allowing, therefore, normalization to function as a flowing and non conflictive "state." This situation results in an exacerbation of what Marcuse argued about societies of the past century: "This means that the individuals are not set off from each other under their own proper needs and faculties but rather by their place and function in the pre given social division of labour and pleasure"

(Marcuse, 2001: 130)

In this direction, we can understand how social normalization is a consequence, but at the same time a generator, of repetition, in the times of social bearability mechanisms and devices for the regulations of sensations. Now, to accurately characterize the "state" of such societies, it becomes necessary to emphasize the experience of immediate enjoyment as a privileged axis through which the elaboration of possible normalizations crosses.

One of the most basic processes in the 21st century structuration process is the connection between Immediate Enjoyment and Consumption that impact at the centre of sensibilities. Some time ago Marcuse elaborated the connections between contingencies, consumption, and enjoyment:

It is not a paradox that the producer recedes more and more before the consumer, nor that the will to produce weakens before the impatience of consumption for which the acquisition of the things produced is less important than the enjoyment of things living.

(Marcuse, 2001: 148)

In direct connection with what we have pointed out regarding normalization related to compulsive repetition, immediate enjoyment is the device through which the diverse and multiple means of generating substitutes, replacements, and satisfactions are updated through consumption, understood as the mechanism to reduce anxiety. The connection between consumption, enjoyment, and objects acquires the procedural structure of addictions: there is an object that liberates moments of containment/adequacy to a specific sensibility state, with such a power/capacity that its absence demands an immediate replacement/reproduction. Without those objects, there is a fracture in the always undetermined emotional weaves, in such a way that the experience of lack induces/produces the need again and the immediate consumption of the referred object.

In this sense, enjoyment can be understood as the complex and contingent result lived as a parenthesis "here-now," as a continuity in time that also produces a state of subjective disembedding. Enjoyment is resolved in the instant as a time/space of realization, which updates without any mediation with the perception of continuity/discontinuity. So, it is an immediate, a "now" that makes sense in its indefinite repetition, by which we can understand why it experiments "in itself" as a continuous flux of time. Enjoyment is the micro/macro marker of hours, days, and years; therefore, it becomes the parameter for the "loss of sense of age." Immediate enjoyment is coupled to the structure of disembedding time/space of societies producing a subjective unanchoring. This means that neither the copresence, nor "face to face work," nor strategies for sheltering subjectivity are (and cannot be) included in the act of enjoyment. That is why enjoyment becomes circumstantial, contingent, brief but "absolute" and radically "here-now."

Immediate enjoyment is an act with totalizing pretensions that suspends the flux of everyday life. Therefore, it is "made," it is produced, performed, and dramatized. Immediate enjoyment refers to a form of "intense" and "superficial" appropriation, restorative of objects for decreasing anxiety through salvific technologies. Immediate enjoyment happens in the moment of consumption, as they are practices with a totalizing pretension by and through which the individual subjectivizes the object, reconstructing it in its structuring potency of vicarious experiences.

Enjoyment being an act, and consumption an action, the dialectics of their mutual interactions define life as a set of practices oriented towards them, with the promise of operating as "erasers" of remembrance of the effort. There occurs in the current contexts a rupture/continuity with what Baudrillard would observe:

> In that level of «experience», consumption transforms the maximum exclusion of the world (real, social, historical) into the maximum index of security. Consumption points to that happiness by a defect which is the resolution of the tensions. But it faces a contradiction: the contradiction between the passivity this new system of values implies, and the norms of a social moral that, essentially, continues being that of will, action, efficiency and sacrifice. Therefore, the intense guilt, this new style of hedonist conduct and urgency carries, clearly defined by the «desire strategists», of un-blaming passivity.
>
> (Baudrillard, 2009: 17)

Today the whole system of beliefs has reabsorbed the content of the sacrificial (such as we analysed in the last section), emphasizing the moment of enjoyment as the act that "makes sense" of consumption actions, series of acts/

actions that materialize in what the "old" society of consumption had turned into a sign.

It doubles the obligation/precept/mandate of enjoyment for the ritual of consumption, as social forms of synthesis that turn an individual appropriation of enjoyment into "the" privileged connection to the social totality. Once again it hyperbolizes what Baudrillard maintained:

> ... the consumer man is considered obligated to enjoy, as an enjoyment and satisfaction undertaking. He considers himself obligated to be happy, to be in love, to be flattered/flatterer, seductor/seduced, participant, euphoric and dynamic. It is the maximization principle of existence by using the multi-plication of contacts, relations, through the intensive employment of signs, objects, through the systematic exploitation of all enjoyment possibilities.
>
> (Baudrillard, 2009: 83)

Consumption becomes paradoxically a "here-forever" that installs with the promise of containing the set of secular parousia, whose technological structures give enjoyment a salvific character.

Consumption contains the keys to heaven on earth, by which the structures of expropriation/depredation/dispossession become secondary and diluted in the promises of total experiences, turning into the materiality that describes the grammar of current class struggle.

Consumption inverts/modifies the connections between objects/individuals, individuals/individuals, and objects/objects, transforming such relations by summarizing them as quantity/quality; volume/intensity; access/denial with another, with "someone"/oneself in and for immediate enjoyment. This way it produces the structuration of the living, of life, and of what is liveable, through consumption that grants enjoyment: consumption becomes a belief.

According to what we have presented above regarding normalization, immediate enjoyment in and through consumption produces derealization as a loss of "contact with reality," remoteness from the adequacy patterns of action in copresence and repressive de-sublimation.

Immediate enjoyment "in-consumption" – being a surrogate strategy of social synthesis – occupies at least three simultaneous positions in the processes of coordination of action: (a) it is a bridge with others, (b) it is a way to elaborate social presentation of the person, and (c) above all, it shelters the contradiction of being an individual act performed in front of others. The social magic of enjoyment lies in its strength to brake/unite the public/private, it consumes to be seen consuming, and it goes as far as the paroxysm of enjoyment if it is dramatized for someone.

In current times, consumption as a central part of "Economy" operates at the centre of the contradictions of capitalist life: in the heart of the dialectics between commodification–de-commodification, in the redefinition between what is private and public, and the restructuration of the experiences of producer/consumers.

Enjoyment in consumption is strongly related to Life Politics (*sensu* Giddens) considering that from that viewpoint it is an answer to the question: what to do with identity? Having mimesis as a goal, the externalizations of the subjects by and in the object turn into a matter to be shown/watched. It is by this path that two convergent processes produce: (a) the redefinition of what is lived as inner experiences, while housed in the circumstantial, in the undetermined, in the contingent, and transversed in the instantaneous, ephemeral, perishable, and denied by excess of ideas of Illouz about the existence of frozen intimacies; and (b) the reconstruction of a social place that is called "intimate," transforming it in the edge between showing/hiding, between the as if/so how, between outside/inside, and between public/private, operating as an inverse configuration emotional de-coercion/ emotional coercion (*sensu* Elias) as a "career" of a disposed of intimacy.

The social ways of "being-in-the-world" find in consumption/enjoyment/ intimacy their marking line, the criteria of validity of what should be considered a life lived intensely. Enjoyment as an "existential" of capitalist life in the sense of experimentation to be told/lived in front of and for others is connected with the daydream states where consumption explains the belief in a world lived to be seen. The belief in mimetic consumption is an experience that structures everybody's lives around showing, as the surface of an inscription of all sensibilities that long for some degree of "veracity." Existence and spectacle join with and through immediate enjoyment.

However, as we have maintained above, the constitution of normalized societies in immediate enjoyment is the result of, and at the same time produces, the intensification of consumption as a strategy of the political economy in many of the states of the Global South and North. One of the privileged routes for "reinforcing"/building the representational situation for Others that consumption implies, is the presence of government and private spectacles as moments of condensation of sensibilities. In this framework, it is necessary to synthesize the three features of the politics of sensibilities that accompany the colonization of the inner planet: The LoW, the elaboration of a PoP, and the Practices of BoG.

1.3.2 The components of politics of sensibilities on normalized societies

1.3.2.1 The logic of waste

Our intention here is to propose a metonymic game: the analysis of waste is an indicator of the "form" of a predatory and sacrificial society. It is in the indicated sense that we want to review here the classification of the discarded in sacrificial societies as a "pre-ludic" to a reconceptualization of the marginal. Those who "treat-with-the-waste" are a testimony of the place of those remains left by a society that predates, are traces of the votive vows left on the sacrificial altar of enjoyment, and are secular offerings so that consumption is perpetuated.

The connection we intend to show is the following: a society normalized in immediate enjoyment through consumption (Scribano, 2013a; Scribano and De Sena, 2014) in the context of predation as a systemic reproduction structure (Scribano, 2012), neocolonial religion as a content of the political economy of morals (Scribano, 2012, 2013b), and the spectacularization of the social (Scribano and Moshe, 2014), together establish as a logic of social interrelation to the practices of waste.

We propose to understand this "extension" of the "practices of scrapping," following the footsteps of the analysis that both Weber and Adorno and Hork-heimer carried out regarding the place of instrumental rationality as an expansion/incorporation of the logic of modern science as a structure of social practices at the dawn of the 20th century.

In the context of the above, what we propose is: (a) to understand that the mode/form of normalization is connected with the styles of immediate enjoyment that are created in "tension" with the "ways of consuming," generating a diversity of "practices of discard;" and (b) understand how the aforementioned practices are one of the basic axes for the constitution of the sacrificial spectacularity that characterizes normalization. To put it another way, discarding has become an "action scheme" and a vector of the practices of feeling that occur as a result of the connections: consumption = waste = predation. It is in this sense that, exploring consumption, sacrificiality, and scrapping, we propose some clues to understanding a phenomenon that we consider central in the contemporary structuring of our society.

In this context we have chosen the following argumentative strategy: (a) we revisit the connections between consumption and enjoyment as features of the normalization of society; (b) we schematize some of the relationships between sacrifice, consumption, and waste; and (c) we synthesize some axes to conceptualize disposal practices.

The normalization of society in and through sacrificial spectacularity implies among other components: the ritualization of putting oneself in the hands of tomorrow as a fantasy of redemption; connections between waste/classification as a metamorphosis of the inequality system; and practices of feeling moulded from the disposable and discarded as interaction. This implies the structure of social relations in a waste society.

In a predatory world, the interaction logics are constituted between the elliptical torsions that are instantiated between the acts of consuming and discarding. One of the cunnings of the current political economy of morality is to extend the practices and gestures of planetary depredation of energy to everyday life. From the beginning of the 20th century, instrumental logic (*sensu* Horkheimer) extended as practical logic (*sensu* Bourdieu); today the predatory logic marks the pre-reflective frames of consciousness and interpretation schemes (*sensu* Giddens) in the form of disposal practices.

Consuming as the inaugural act of sacrificial capitalism implies (in this argumentative context) the extension of its "format," encompassing the connections of human beings with objects, with all living beings and with themselves. Every act of consumption involves waste, scraps, and rubbish. Three results, but at the same time conditions of possibility of any relationship in the world of the disposable, that in their volume and diversity act as models of the "next" actions of enjoyment-in-the-consumption.

Discarding thus involves the part of the object/subject/process that is left unconsumed, given the structure of the act itself and the part of the object that, given the condition/structure/materiality of the object, it is inappropriate/impossible to destroy.

The sacrificial is stressed with and disposable as the practical logic of doing in contemporary capitalist society. A society that takes waste to use as merchandise, a society that sacrifices what is on the sidelines, a society that delivers a daily offering on the altar of production so that the world continues as it is, a society that ritualizes neocolonially bodily energies in pursuit of predation, structures a set of social relationships around the triad consumption = object = dispossession.

The objects stand upside down and begin to rule the world of human being in and through a fetishism that made body incarnate the practices of a secular religion valued in the political economy of morality. The dialectical game between consuming, discarding, and enjoying opens a reorganization/restructuring inter and intraclass, and repeats the extension of the conditions of possibility of capitalism as a system.

The waste is a point of the sacrificial strategy of a society of compensatory consumption: normalized in the immediate enjoyment the society operates a double substitution, the offering becomes new waste and the remains metamorphose into offerings. The substitutive efficacy of consumption rewrites what there is of the person represented in the propitiatory victim; it is the represented person who operates in mimetic identity with what is offered (object), the naturalization of his victimization to maintain order/consumption and remove the shadow of the crisis. To chase away the crisis by transposing the object of the object is the role of waste in the waste of your life, it is a life so that life-in-consumption does not cease. Where there was a waste, consumption remains and through it, the reproduction of enjoyment is obtained on the altars of the compensations to dissolve the conflict. Bypassing the conflict by concentrating on what is sacrificial in the offerings of the disposable indicates the "invention" of "external power" to the human being that makes the scraps a "logical consequence" of a society that depredates and despoiled.

In many of the ways indicated here in the studies that make up this book, we believe are important clues to better understand the world of waste in terms of its central agents. That is why analysing waste is essential for exposing how the apparently neat margins, edges and boundaries drawn in cities in fact serve to hide the stink of the effects of immediate enjoyment in today's society.

1.3.2.2 Politics of perversion (PoP)

George Orwell, in his novel *1984* (1949), paints a world in which the war agency is called the "Ministry of Peace." The agency responsible for manipulation, propaganda, and historical falsification is called the "Ministry of Truth." And the one which perpetrates punishments and torture is called the "Ministry of Love." In a similar description – metaphorically speaking – we want to inscribe the conceptualization of a PoP, in a sense very close to that analysed by Adorno (2008), linking political and aesthetic categories to describe the "fit" individual/world, arguing that his description is triumphantly achieved in *Brave New World* (1932), by Aldous Huxley. Hence, in the contact nodes between *1984* and *Brave New World* we intend to outline the conceptualization of the existence of a PoP.

Returning to some of the clues already outlined, the PoP consists of making lies, manipulation, fictionalization, etc. a desirable "state of affairs" as a central strategy for managing emotions. We argue that we are facing policies, since the actions referred to can be better understood if they are inscribed in what we have called "politics of bodies," "politics of emotions," and "politics of the senses." We affirm that they are political because their modulation and execution must be thought in terms of the patterns of feeling that are elaborated in the tensions between sociabilities, experientialities, and sensibilities. Managing sensations, managing emotions, and designing sensibilities, a set of practices of feeling that shape the political economy of morality, is outlined. It is precisely in the pretence of directing the social cement that is found/elaborated/arises between ethics, morals, and aesthetics that there emerges a politics whose main objective is that the auditoriums feel and that the subjects participate in the spectacle and the sacrificiality.

But these politics have a particularity: they are actions whose object/effect is to camouflage, to pass off "one-thing-for-another," to deny what exists, and to deny what exists by an exhibition, by demonstration, and by exuberance. They are perverse practices; practices of investing completely, emphasizing the reverse. These are state practices, government efforts, and the unintended consequences of state action.

The politic of perversion consists in the fact that the paradoxical link is established as the axis of a sensibilities management strategy. It is a way of refusing the real; it is to not relocate (be) in what is manifested; it is blocking what bothers. It is a way of splitting the acting self: "the-who-does-practice" unfolds and splits to make (be) bear what is in the break. The perversion of politics is the unnoticed acceptance of the renunciation of social change ("there will always be poor people"). It is an emphasis on self-satisfaction (it is what "I feel"), it is a form of "acceptance of evil" that seeks to expand, reproduce, and massify. It is the epic hyperbolization of the acceptance that "there will never be anything different." A politics of perversion is identified when one thing is said for another, when it shows what it is not, when it appears, when it is simulated, the sensibility of others is manipulated/managed when there is a policy of shows intended to hide. Politics is a perversion as an exercise of a ministry (*sensu* Bourdieu).

1.3.2.3 Banalization of good (BoG)

The BoG has at least six basic characteristics that are threaded into two pairs of elliptical and dialectically arranged triads: 1) fetish-dogma-heteronomy and 2) epic-gesture-narration. These triads are registered and displayed in an irregular space qualified by four complementary nodes of resignation, as a component of the neocolonial religion: fake contention/dependence/romanticism /miserabilism.

The two triads are, in short, moments of the same helical movement where each of its moments comes into tension with the other and with the passing at different times through the same place, but in a different "state of affairs." Thus fetish-dogma-heteronomy and epic-gesture-narration integrate the features of the BoG and are located on an inscription surface constituted/drawn within the framework of four points that interact geometrically: fake contention/dependence/romanticism /miserabilism.

The features of the BoG are assessed enhanced and better understood if it is possible, at least partially, to investigate how it is produced and/or what processes it implies. The procedure is simple. The contents and senses are selected whose acceptance and prestige are part of: (a) a common foundational story, (b) involve characters of mythical valence, and (c) actions associated with heroic deeds. These objects, processes, and characters are placed at the service of action and/or events that have a particular character with claims of universalization; they are processed. They enrol in productions of "benevolent" results that, when liquefying the original emotional evocation, truly empty the truth. When it is aestheticized and massified, what invoked the moral force of goodness collapses and is reduced to a mere process.

At present, the BoG can be observed in three aspects of aesthetics as a political phenomenon: the dismantling of the contents referred to as "the revolutionary," the loss/loss of a "happy ending" for history, and the reinvention of the collective resolved in terms of a "new" individualization.

It is in this framework of the BoG processes that the place of fiction emerges strongly. Although "fictionalizing" and "pretending" are not equivalent practices, they go through the "procedural" from three axes: (a) an "as if" society, where the really important thing is to represent/dramatize the experience, not to live it; (b) a society where "having strong/deep experiences" is an imperative of the political economy of morality; and (c) the hyperbolization of the appearance as pornographic diminished humanism.

Having an experience and living an experience becomes, on many occasions, the practice of buying such experience or of "making-of-account-that-I have lived." The BoG is an imposter in its metaphorical transfer of controlling and manipulating a state of affairs to appear, to emerge. It is not necessary to live, it is enough to show that I have lived, and it is not necessary to believe, only to show that I believe.

The society standardized in enjoyment through consumption, and the three components of the politics of the sensibilities that characterize it and are just enunciated, are the space where the structuring of Society 4.0 is inscribed and its consequences for the colonization of the inner planet are manifested.

1.4 Society 4.0 and sensibilities

This section aims to offer an introductory and schematic overview in which the current contents of the research on Society 4.0 and the politics of the sensibilities are exposed. The section seeks to make observable how the spheres of Society 4.0 interrelate with the features of digital labour in such a way as to generate the current politics of sensibilities. In the search for connections between the transformations in work and society, the chapter ends by suggesting the idea of the emergence of "sensibilities of platforms."

In this Society 4.0 there is an important transformation of the political economy of morality, the politics of sensibilities, and the political economy of truth associated with it: the structure of the political economy of morality accompanies changes in the economic politics of truth. It modifies the set of accepted processes to produce the truth, the criteria to accept perception as true, and the specialized areas to "guarantee" the truth. The processes to obtain the truth move away from the traditional scientific procedures that move towards logics articulated around sensibility and emotion. From the various forms of empathy, perception, and sensations of "grasping," through intelligent regimes of emotional regulation towards alternative spiritualities, they intersect and are articulated as possible ways of reaching the truth.

Socioneurology, informative diagramming of haptic systems, fuzzy logic, body/machine interfaces, nanotechnology, genetic design, and artificial intelligence are some of the scientific procedures of the 20th century that "help" new ways of obtaining the truth.

The predatory expansion of capitalism on a planetary scale has generated a rapid, complex, and massive articulation between the features of so-called Society 4.0, digital labour, and a political economy of morality. Some of the features of the transformations taking place in this current social structuration process include the expansion of the revolution 4.0 and its impact on productivity and labour, the massification of a political economy of morality based on non-truth, the growing number of refugees and migrants around the world, military tensions, and wars of a multilateral nature.

The modifications in work and its consequences in the social structure are central axes of the history of humanity: the crossing between production, needs, goods, models of work, organization, and wealth distribution systems have been, are, and will be the constituent axes of societal forms. In the same vein, it is possible to understand how technological transformations have involved modifications in work and in social relations as a whole. These technological changes imply, in one way or another, variations in the mode by which people relate to time, space, shortage, and satisfaction.

In the described context it is easy to understand how and why the expansion of digital labour comes about in the context of the massification of the modifications produced by the digitalization of society, generating consequences in the politics of sensibilities. The digital society brings together the expansion of Industry 4.0 (and digital labour) with the widespread globalization of digital consumption. It is in this intersection/convergence that "new/diverse" features of the politics of sensibilities are elaborated.

To paraphrase what Montesquieu said about the connection between trade and capitalism through digital commerce and mercantilization, consuming is softened and sweetened. The politics of digital sensibility promises instant consumption without conflict, taking for granted the thousands of people "behind the scenes" that make those sensations possible.

As we have already argued above, it is in this scenario that the planet is experiencing a process of social metamorphosis on a global scale. One of the consequences and central effects of the relations between Society 4.0, digital labour, and the social structuration process are the changes in the politics of sensibilities. And if we know that these politics "are understood as the set of cognitive-affective social practices tending to the production, management and reproduction of horizons of action, disposition and cognition" (Scribano, 2017: 244), then to reflect on the changes mentioned is essential to understand the current situation of society.

In this context, we detect that in the existing literature there is a lack of attention to and research about how the politics of sensibilities are being altered by the current situation of Society 4.0, given that digital labour implies substantial changes in the lives of individuals, groups and society in general. Just as an example, this implies that we also have to analyse how the situation of the geometries of the bodies, the grammar of actions, the politics of the bodies/emotions in relation to what is the digital era, digital platforms, teleworking, etc., now modify day-to-day life.

Every form of work, and especially digital work, involves certain ways of managing at least two spheres of the world: those connected with the senses (hearing, touching, tasting, looking, and smelling), and those that articulate perceptions, sensations, and emotions. The politics of sensibilities are developed from, among other factors, the states of these two spheres.

The mobile/digital revolution implies transformations in the management of labour and vice versa, and under the umbrella of these modifications are developing new politics of sensibilities. One of the most important aspects of the advent of companies 4.0 is the rapid development of social networks and the enormous growth of their commercialization and commercial value. In this framework, the interactions between the face-to-face social world, the virtual world, and the "mobile" world of cell phones and tablets have grown exponentially.

Many authors argue that we are facing the Fourth Industrial Revolution, and that this can be characterized by the consolidation of at least three factors: (a) the appearance of Big Data as a resource for social diagnosis, (b) the Gig Economy as evidence of the growth of deinstitutionalization, and (c) the Internet of Things (IoT) as a new form of production and "management of sensibilities."

For its part, the use of Big Data analysis implies material surveillance of massive amounts of information about people and societies; the Internet, social networks, and mobile interaction as a space for searching, construction, management, and distribution of information; the digital dependence of the most dynamic sectors of the "real" economy; changes in the management of work and appropriation of the benefits of capital; and the intimate relationship between the depredation of environmental assets and computer/digital assets.

In relation to the Gig Economy, it is possible to notice as central features: flexibility in the modalities of coordination of action, transformations in resources to guarantee competences, the contingency of temporary and spatial links between the consumer and the producer, and change of the means of payment for services and goods.

On the other hand, the IoT brings with it the following consequences: a new kind of "do it yourself" paradigm; redefinitions of proximity/distance between the product and the producer; and modification in the relationship between "materials"/sensation.

The increasingly important weight of "The Cloud" as a virtual space for production, storage, management, and distribution of information must be added. Indeed, among the many factors that converge for the modification of the modes of work management, knowledge, and production at present, the Cloud is the most important one. This is so, since (a) it is a virtual space designed to improve collaborative work, (b) it allows the obviation of the inequalities of access to expensive hardware, and (c) it enables the promotion of a more "agile" information management.

Another feature of the connection between Society 4.0 and labour is the so-called "sharing economy," as maintained by Parente and his colleagues:

> The popularized "sharing economy" term has been used frequently to describe different organizations that connect users/renters and owner/providers through consumer-to-consumer (C2C) (e.g., Uber and Airbnb) or business-to-consumer (B2C) platforms, allowing rentals in more flexible, social interactive terms (e.g., Zipcar and WeWork).
>
> (Parente et al., 2018: 53)

> 'Collaborative consumption' and the 'access-based economy' are other ways to identify a set of economics interactions based on the Internet as a platform and implies countless social transformations. One of these changes is the new and stronger role of consumption and consumers in shaping economic interactions. In some way, these interactions modify the practices of having, possessing and using under the influence of the Internet and the time-space resignification that this implies. The perceptions about what it means to be an owner are confronted by the 'experiences of using.'
>
> (Davidson *et al.* 2017)

In the wake of the global financial crisis that began in 2008, consumers sought other means of accessing products and services, aside from property charges. A new economic model emerged, known as the sharing economy or collaborative consumption, which integrated collaboration, technology, and the desire to be more effective.

Society 4.0 is an increasingly clear and forceful reality, an indicator of this is the sustained growth of electronic commerce and the predictions that indicate the same path for the coming years. In the recent eMarketer report coordinated by Andrew Lipsman (2019) it is very transparent about the predictions:

> Recently also (2019) a global study by Gallup for Wellcome Global Monitor on the perception about science, health and personnel of health and science, the high degree of confidence in science by a significant proportion of subjects in the world can be perceived: "Worldwide, about seven in ten

people feel that science benefits them - but only around four in ten think it benefits most people in their country. (…) About a third of people in North and Southern Africa, and Central and South America feel excluded from the benefits of science. South America has the highest proportion of people who believe that science neither benefits them personally nor society as a whole, about a quarter of people."

(Gallup, 2019, in Lipsman, 2019: n/p)

In the aforementioned context, trust emerges as a central problem for 4.0 societies in various planes and aspects, but primarily under the cover of cybersecurity and security in the use of personal data; the processes intersect and interact with the expansion of the redefinition of institutions, as in the case of Uber, where one can observe a flexibilization, "liberalization," and resignification of state controls. In the same direction, it is possible to verify the growing impact of Artificial Intelligence in its different applications such as chatbots, robots, drones, and other objects/processes linked to the IoT and the modifications that this involves for everyday life.

In this Society 4.0, there is an important transformation of the political economy of morality, the politics of sensibilities, and the political economy of the truth associated with it. The structure of the political economy of morality accompanies the changes in the political economy of truth. It modifies the set of processes accepted to produce the truth, the criteria to accept perception as true, and the specialized areas to "guarantee" the truth.

As instruments designed to "see things happening," drones "stretch" the current paradigm of sensibility, playing the role of witnesses that reproduce reality from a distance.

What is sustained up to now is more complex if we bear in mind that Society 4.0, among many other things, has transformed the potential of communicating through photographs, videos, and audio recordings expressed in terms of cell phones and smartphones. What also brings its use as a technique to record, portray, and interpret the world (Lansen and Garcia, 2015).

From the different forms of empathy, perception, and sensations of "capture," through intelligent regimes of emotional regulation, to alternative spiritualities, all these are crossed and articulated as possible ways of reaching "the truth" and/or "new sensibilities states."

Precariousness, massification, instantaneity, and digitization of daily life are practices that consolidate the era of a new "politics of touching" and diverse "politics of seeing," and work is a space where these processes are rapidly and strongly evident. In the scenery of the globalization of societies normalized in immediate enjoyment through consumption, the processes of classification and qualification of touch are modified, thus a renewed "politics of touching" assembles different (and diverse) geometries of the bodies deployed in virtual environments and by digital resources. In the same vein the proximity and distance between gazing, seeing, and observing are transformed on the digital horizon. From 3D effects, passing through augmented reality, to the arrival of drones in daily life, we see modification and the appearance of a new form of "politics of seeing." Both the new "politics of touching" and diverse "politics of seeing" have a central impact in redefining the "world of work:" "new" environments, resources, processes, and goals have been created and with them, the labour and workforce are transformed.

1.5 Politics of sensibilities and sensibilities of platform

In a study "suggestively" called "Skills for Social Progress. The Power of Social and Emotional Skills," the OECD links emotional skills and development, revealing that emotions are objects of intervention on the part of states and the market. As "skills beget skills," early interventions in social and emotional skills can play an important role in efficiently raising skills and reducing educational, labour market, and social disparities (OECD, 2015: 14).

In this framework, it is possible to understand how the digitalization of the world coexists with the emotionalization of the processes of domination and of everyday life. We are facing a social system that is globalized by producing/buying/ selling emotions in and through the media, social networks and the Internet. In this sense, it is probable to understand what it is possible to call the emergence of "sensibilities of platform."

There are many sides that interweave possibilities and limitations in some renewed ways to release/repress the creative energy of human beings, and are just some examples of how the inhabitants of this planet are living in the 21st century. These are some of the possible paths to investigate the connections between working and society, which are nowadays involved in the battle of sensibilities.

We live in a virtual/digitally connected world shaped by the technological transformations of the last 10 years. The Internet and mobile telephony are two vectors that set the stage for three significant changes in the politics of the sensibilities: (a) the organization of the day/night is unlinked from the experience of the subjects that experience it, (b) the modification of the sensations of classification, and (c) valuations on world modifications.

Each society has a preponderant way of managing work and this constitutes a central axis of the politics of sensibilities. Society 4.0 implies the massification of digital labour and with it the "sensibilities of platform." A "sensibilities of platform" emerges in this Society 4.0 that is immediate in three senses: (a) in the vehicle in which the action resides (it is the feeling of always being "on line"); (b) it is a society that "is during use," "between," "in passing;" and (c) is pure presentification (here/now). Much of digital labour has the same characteristics and, in this direction, the political economy of morality consecrates this way of "feeling the world."

It is from the "immediate" that appear some of the connections between the digitalization of society and the consolidation processes of societies normalized in immediate enjoyment through consumption. The ephemerality of enjoyment resembles what many authors associate with digital labour as disruptive or creative destruction. The immediate is similar to an "on-demand" platform strategy.

In this context, perhaps "in the course," "in this action becoming permanently," in this idea of the immediate and ephemeral, we should interpellate the world with silence. An act of listening where one feels the other in a different way. Silence is the starting point of dialogue as a matrix of knowledge and life become personal interaction.

In this context, "to be always moving," "in the course," "in this action to become permanently," within the framework of the perception of the instantaneous, immediate, and ephemeral, digital consumption develops.

Thus, the complexity of our days could be characterized by the expansion of societies with standardized practices of immediate enjoyment through consumption where a phenomenon of internationalization of emotionalization (Scribano, 2017) is extended, which runs through public policies, processes of discrimination and work. On the basis of these axes, it is possible to reconstruct new geometries where the grammars of actions are associated with the game of proximities and distances that redefine the subjects' own constitution, their associations (collectives), and the State itself (Scribano, 2015).

This image, made up of at least some of the brushstrokes that we can briefly reconstruct here, becomes an interesting starting point to reflect upon the knowledge processes structured as condition/consequence of this scenario. For the perspective we would like to develop here, it is especially important to highlight the connections between epistemology and emotions (Scribano, 2012).

One of the possible ways of scrutinizing these connections involves exploring the relationship among "word," "knowledge," and "sensibilities" in phenomena that cut across the contemporary lifeworld such as "work 4.0." It becomes evident that "virtual spaces" and, consequently, our work experience become accessible to us "in" and "through" specific politics of sense: that is to say, "adequate" ways of touching, seeing, and hearing.

Knowing becomes an extension of the sensibilities that are globally commodified. In the world 4.0, expert knowledge, the scientist, and knowledge about everyday life intersect and intertwine in daily experiences. An epistemology of social sciences that tries to improve the capacities to know the world should incorporate the different modalities of reflexivity that make up the diverse modulations between knowing and sensing. Thus, the relationship between truth and language must be reconnected beyond that dualism in the framework of a dialectics between knowing, sensing, and saying.

The colonization of the inner planet is complex process that involves the tension between, (a) a "new neocolonial religion" whose dogmas of faith are mimetic consumption, resignation, and diminished humanism; (b) a normalization that includes the internationalization of emotionalization, of the LoW, the PoP, and the BoG; and (c) a "sensibilities of platform" that arises from 4.0 society that is instantaneous in three senses, in that the action resides in the vehicle, it is a society that "is in use," "between," "in passing," and it is pure here/now.

In the next chapter, it becomes clear how society becomes a body invading it, occupying it, and micromanaging it, this being one of the two central processes by which the three diagnoses that have been completed become materially body.

Note

1 We approach here, with several differences, to E. Fromm in his exposition of the idea of industrial religion. CFR Fromm (1977).

References

Adorno, T.W. (2008) *Crítica de la Cultura y Sociedad I.* Madrid: Akal.S.

Baudrillard, J. (2009) *The Consumer Society.* UK: SAGE Publications.

Davidson, A., Reza Habibi, M., and Larochec, M. (2017) "Materialism and the sharing economy: A cross-cultural study of American and Indian consumers", *Journal of Business Research*, 82, (2018), pp. 364–372.

Huxley, A. (1932) *Brave New World.* UK: Chatto & Windus.

Marcuse, H. 2001 (1968) "The movement in a new era of repression." In: Kellner, D. (Ed.) *Collected Papers of Herbert Marcuse Volume Three.* UK: Routledge.

Lansen, A., and Garcia, A. (2015) "But I haven't got a body to show': Self-pornification and male mixed feelings in digitally mediated seduction practices", *Sexualities*, 18(5/6), pp. 714–730.

Lipsman (2019) "Global Ecommerce 2019", Report Collection, Jun 27, *EMarketer* https://www.emarketer.com/content/global-ecommerce-2019

OECD (2015) *Skills for Social Progress: The Power of Social and Emotional Skills.* Paris: OECD Skills Studies, OECD Publishing, https://doi.org/10.1787/9789264226159-en

Orwell, G. (1949) 1984. *Animal Farm: A Fairy Story* London: Secker & Warburg.

Parente, R.C. et al. (2018) "The sharing economy globalization phenomenon: A research", *Agenda Journal of International Management*, 24, (2018), pp. 52–64.

Scribano, A. (2009a) "Ciudad de mis sueños: hacia una hipótesis sobre el lugar de los sueños en las políticas de las emociones." In: Lavstein, A. and Boito, M.E. (Comp.) *De insomnios y vigilias en el espacio urbano cordobés. Lectura sobre Ciudad de mis sueños.* Córdoba: Sarmiento editor.

Scribano, A. (2009b) "¿Por qué una mirada sociológica de los cuerpos y las emociones? A Modo de Epílogo." In: Scribano, A. and Figari, C. (Comp.) *Cuerpo(s), Subjetividad(es) y Conflicto(s) Hacia una sociología de los cuerpos y las emociones desde Latinoamérica.* Buenos Aires: CLACSO-CICCUS.

Scribano, A. (2010) "Cuerpo, Emociones y Teoría Social Clásica. Hacia una sociología del conocimiento de los estudios sociales sobre los cuerpos y las emociones." In: Grosso, J.L. and Boito, M.E. (Comp.) *Cuerpos y Emociones desde América Latina*, pp. 15–38. Buenos Aires: CEA–CONICET.

Scribano, A. (2012) "Sociología de los cuerpos/emociones". In: *Revista Latinoamericana de Estudios sobre Cuerpos, Emociones y Sociedad - RELACES.* N°10. Año 4. Diciembre 2012-marzo de 2013. Córdoba. ISSN: 1852.8759. pp. 93–113. Disponible en: http://www.relaces.com.ar/index.php/relaces/article/view/224

Scribano, A. (2013a) *Teoría Social, Cuerpos y Emociones.* Buenos Aires: Estudios Sociológicos Editora.

Scribano, A. (2013b) "Una Sociología de los cuerpos y la emociones desde Carlos Marx." In Scribano, A. (comp.), *Teoría Social, cuerpos y emociones.* Buenos Aires: Estudios Sociológicos.

Scribano, A. (2015) *¡Disfrútalo! Una aproximación a la economía política de la moral desde el consumo.* Buenos Aires: Elaleph.com.

Scribano, A. (2017) *Normalization, Enjoyment and Bodies/Emotions: Argentine Sensibilities.* New York: Nova Science Publishers.

Scribano, A. (2018) *Politics and Emotions.* Houston USA: Studium Press llc

Scribano, A. (2019) *Love as a Collective Action: Latin America, Emotions and Interstitial Practices.* UK: Routledge.

Scribano, A., De Sena, A. (2014) "Consumo compensatorio: ¿Una nueva forma de construir sensibilidades desde el Estado?", *Revista Latinoamericana de Estudios sobre Cuerpos, Emociones y Sociedad.* N°15. Año 6. Agosto 2014 - Noviembre 2014. Argentina. ISSN: 1852-8759. pp. 65–82.

Scribano, A. and Lisdero, P. (Eds.) (2019) *Digital Labour, Society and Politics of Sensibilities.* UK: Palgrave Macmillan.

Scribano, A. and Moshe, M. (2014) "Spectacles for everyone: Emotions and politics in Argentina, 2010–2013." In Moshe, M. (Ed.) *The Emotions Industry*, pp. 161–180. New York: Nova Science Publishers.

2 Colonization of the inner planet I
Body reconstruction

2.1 Introduction: The colonies of the 21st century: sensations, emotions, and sensibilities

As the former Prime Minister of Spain Rodríguez Zapatero stated at the inauguration of the International Iberian Nanotechnology Laboratory:

> Applying the most advanced technologies to the solution of so many problems that humanity has in many fields is also a matter of our time that we have to face. In the age of the ancient navigators, Portugal and Spain stood out in the field of geographical discoveries. With this centre, we now express the will to contribute to drawing the new atlas of the future.
>
> (La Nación, 2009)

Colonizing is occupying, appropriating, and managing; today none of these actions can be carried out without the help of science and in this multidisciplinary configuration the colonization of the inner planet finds one of its central axes. Developing a critique of this connection between science and the elaboration of a new colony is the challenge that we must accept in a world where there is a powerful expansion of commodification of the production of sensations.

21st century capitalism has started another trip around the world. These are not now the journeys of Marco Polo, Colón, or Magellan – it is the entire edifice of science that is navigating an even more unknown world, one where much of the wealth of the future lies: our bodies/emotions.

The journey is long, it has just begun, the surface is still unknown, and the new silk is the sensibilities, but it is absolutely important to emphasize the epigenetic results of what will be developed here, since as Hoffmann and his colleagues argue that

> … the concept of molecular epigenetic switches illuminates the catalyzing function of epigenetic mechanisms in the mediation between dynamically changing environments and the static genetic blueprint. In the context of Waddington's epigenetic landscapes of valleys and mountains, molecular epigenetic switches are guiding dynamic changes in gene expression underlying robust changes in cellular and organismal phenotypes. Ultimately, molecular epigenetic switches may also serve to reconfigure Waddington's epigenetic landscape for better or for worse by lowering the threshold for transitions between distinct developmental trajectories and increasing flexibility regarding unanticipated challenges. Although of potential benefit, such changes are also likely to increase the risk for human disease.
>
> (Hoffmann, Zimmermann and Spengler, 2015: 12)

And it is in this context of the generational impacts of the colonization of the inner planet that the connection between all human beings, living beings, and the entire metabolic building in which their connections are in place stands out. Here it is important to remember what was stated in the introduction regarding crossings and ruptures between the body-individual, the body-subjective, and the body-social as a way to understand the constitution of the human body.

Beyond the existing theoretical differences, the world of emotions is built with the social foundations of the body: we are what we eat, drink, breathe and in that sense, we are already colonized. Our bodies/senses/emotions are being occupied by transgenic foods, carbon dioxide pollution, flavoured waters and hundreds of products created and tested on a molecular and nanoscale that impact in various ways on our central nervous system, our endocrine system, and the immune system. Hundreds of ships sent by the new navigators seek the gold of the new Indians, take up the new silk routes, traffic new flavours, textures, smells, colours, and sounds; following Montesquieu's advice, they elaborate the bases of a sweet trade between corporations and countries.

As part of the colonization of the inner planet, the colonization of emotions, and colonized emotions, in the 21st century have at least three paths of entry and production:

1. The search for the interpretation of the meaning of emotions with different strategies of identification, reading, and their hermeneutics. The paths used range from surveys, through Artificial Intelligence, until reaching the application of sensors.
2. The construction of devices, robots, and mechanism that reproduce and interpret sensations and cognitive-affective evaluations of emotions and sensibilities.
3. The design, elaboration, and implantation of interfaces, protheses implants, nanorobots, and the establishments of mechanisms, processes, and substances that modify the structure perceptions/sensations/emotions/sensibilities.

As it is possible to see in these alternatives, a complex amalgam between Big Data, the Internet of Things (IoT), and Machine Learning is used and associated with the management of emotions. Each of these paths can be taken independently, or proceed as a process in stages that begins with knowing, continues with "imitating," and ends with inhabiting. As scientific, technological, and business endeavours these paths interact, nurture, and converge.

In this context, it is possible to identify three poles of the colonization of the inner planet that in different phases of progress are reaching the shores of the new world carried by the new ships:

The **first pole** is linked to production and distribution in industry, consumption, and the enjoyment of endocrine-disrupting chemicals (polymers) used in hundreds of products and processes. These substances are found in everything from children's school supplies, through personal deodorants, to the components of cell phones and computers. These substances can "imitate" hormones, making their introduction into the bodies at low doses impact everything from pregnancy to the life of the elderly. Its consequences are among others: the advancement of menarche in girls, decreased sperm fertility, and neurological disorders. "Post-pandemic" science must denounce (as do the Spanish unions and the European Union among others) the use of endocrine disruptors (EDs) as the silent pandemic paradigm.

The **second pole** is the design and genetic manipulation through the use of DNA that is easily identifiable today, encompassing both animal, plant, and human genetic maps and management. The so-called transgenics are genetically modified organisms, that is, those in which the genome has an added or altered gene in their cells. The uses and scope of genetic engineering through which genes can be introduced from one species to another is a highly debated topic today in times of pandemic. The invasion of our bodies has as a link the modifications of the bodies of other species. Taste, texture, smell, and appearance of food, altered repeatedly for decades, transform our senses.

The **third pole** is the use of nanotechnology as a tool for exploration, intervention, and transformation of the body/emotions, constituting a new form of colonization and commercialization. An example is the use of nanoparticles as in the case of cerium oxide, which has extraordinary versatility and is specifically a "rare earth" nanomaterial used for biological applications.

From the three poles and diverse journeys, let us now consider nanotechnology as an obvious path for the colonization of bodies/emotions, and whose consequences are yet to be explored. Just to offer two examples: one of the best known and most developed is the nanotube, and the other is the Internet of Nano Things (IoNT). Carbon nanotubes (CNTs) that are created and used by health sciences end up being a form of occupation and "management of senses." An example is when CNTs as biosensors (especially nonbiological nanosensors) are used as wireless antennas that act as sensors and send the data collected in the body to an external integrator module. As presented by SINlist (SIN – Substitute It Now), CNTs have been shown to be carcinogenic. They induce lung cancer by penetrating lung cells, causing inflammation. They are also persistent and there is evidence that they are toxic to human reproduction.

Another experience can be observed in which Dutta Pramanik and his colleagues have recently reviewed the uses of nanotechnologies, and among others, they point out the properties of the IoNT. In the healthcare context, nano things refer to the miniaturization of biosensors and medical implants at the nanoscale. Intelligent drug delivery and nanoscale surgeries are the flip sides of sensation management.

Obviously, many of the nanotechnological tools of colonization are presented as "advances" of medicine, just as "sea navigation" was presented as a benefit to civilization. The use of cerium is an excellent example, given the potential of being both a clear example of a new human/earth metabolism and successful use of nanotechnology; on the one hand, it is "rare earth" and on the other, it is an excellent nanoscale "vehicle" applicable in various "territories" of the human body. That is, it is of a predatory nature that spreads throughout the body making this an integral part of the geopolitical dispute over "rare earths," and is also a transport mechanism that allows the "occupation" of the body at a cellular level.

As of 2014 Can, Xu and Xiaogang Qu in their review work on the multiple applications of Ceriun advised of its impact as a "drug delivery devices and bioscaffold:"

> Recently, nanoparticles have shown tremendous potential use in biotechnology as drug delivery systems and bioscaffolds. CeONP, a nanomaterial with pharmacological potential, could be used as a nanocarrier or scaffold and also act as a therapeutic agent. Combining the two properties increases CeONP's potential application in biotechnology.
>
> (Can and Xiaogang, 2014: 11)

Recently, since in the framework of the set of metal oxide nanoparticles cerium oxide nanoparticles (nanoceria) produce a great attraction for their varied possibilities of application, Thakur, Manna, and Das (2019) in their paper dedicated to the biomedical applications of nanoceria show its use in antibacterial, antioxidant, anticancer, drug/gene delivery, antidiabetic, and tissue regeneration. In said context they hold that:

> Nanoceria has been found to exert a profound antibacterial effect against different strains of bacteria. The usefulness of nanoceria depends upon its inherent property of showing variable oxidation states, due to which it can act as an excellent antioxidant agent and protect the healthy cells from oxidative stress.
>
> (Thakur, Manna, and Das, 2019: 3)

There are two points we want to make here: versatility and origin. This component, but in reality all nanoparticles, is "usable" in many areas of life through "commercial applications" in varied spaces such as the development of new materials, the pharmaceutical industry, medicine, and environmental technologies. The other side is that being a "rare earth" substance, it is part of the current cold war between the United States of America and China, the latter being the largest producer of these materials in the world.

The ships have arrived and have a gentler and more persuasive appearance than those of Columbus. Unlike 500 years ago, they do not bring "coloured mirrors" but devices that produce sensibilities. They no longer want metal, they want data, and they no longer serve only a single Emperor. A more detailed look at EDs can make it easier to understand how this unnoticed landing campaign is carried out in the bodies.

2.2 Endocrine disruptors as builders of "inner planet"

The consequences of EDs are literally a catastrophe made flesh as they are a set of external factors that impact life and change continuously, but which in turn configure its persistence over time.

Concern about the effects of ED on human life and health is not new in Latin America, as evidenced (just to mention a few) by the works of Argemi, Cianni and Porta (2005), Trossero et al. (2009) and Lerda (2009) in Argentina; Estrada-Arriaga et al. (2013) in Mexico; González and Alfaro Velásquez (2005) in Colombia; and Meyer, Sarcinelli and Moreira (1999) in Brazil. As can be seen, the dates of the works cited here and the plurality of sources, themes, and evaluations indicate the type of impact of the EDs in time and space.

In the same way, actors ranging from the United Nations Environment Program and the World Health Organization (UNEP/WHO, 2013), through the Directorate for the Environment of the European Community (Kortenkamp et al., 2011) and the German Environment Agency, to the Trade Unions, have warned (and produced material on the matter) about the need to regulate ED as harmful to human health. In the same direction, there have been lists recommending substances that should not to be used, or whose use should only be allowed under certain conditions. Also, in the aforementioned context, the epigenetic implications of EDs are known, as argued by Greally and Jacobs in a study originally prepared for the OECD:

> The mechanism by which the group of chemical substances called "endocrine disruptors" exert their phenotypic effects continues to be partially understood, but there is emerging evidence that cell epigenome dysregulation is involved.
>
> (Greally & Jacobs, 2013: 445)

The endocrine system is a complex internal chemical system that regulates vital functions of our body such as reproduction, embryonic development, the immune system, and also cognitive/affective aspects of human beings. The substances that regulate these functions are called hormones.

EDs are natural and/or artificial substances that act like hormones, alter their production and secretion, and interfere with their functions and their elimination. These are chemical substances capable of altering the hormonal system and causing different kinds of damage to the health of exposed women and men as well as their daughters and sons, the latter with more worrying effects in those who are exposed during pregnancy and breastfeeding. They also affect the reproduction and health of other animal species due to environmental pollution. The effects of EDs occur at very low doses, generally well below legally established exposure limits.

Table 2.1 Some Endocrine Disruptor

Bisphenol-A (BPA)	Dioxin	Atrazine	Phthalates
Perchlorate	Retardants Fire	Lead Mercury	Arsenic
Substances perfluorinated chemical (PFCs)	Organophospliate	Esters glycol	Lead

Source: Prepared by the author.

Everyday life is a calamity if you look at the structure of *EDs as pregnancy* in René Thom's terms. Deodorants, body creams, disinfectants, shaving creams, and toys for children are the ways that every day we "impregnate" ourselves with biphenol, phthalates, etc (Table 2.1).

If we explored some possible definitions of ED, it is possible to find clear clues as to how their design, production, and circulation have direct consequences on (and through) human health, the social structuration processes and also for the constitution of bodies.

The Society for Endocrinology (SE), based in the United States, in 2009 produced a first statement regarding the impact and importance of endocrine chemical disruptors for human health. In it, the complex and intergenerational consequences of human exposure to EDs are noted. Thus, the SE follows the definition offered by the North American Environmental Protection Agency, conceptualizing EDs as:

> An exogenous agent that interferes with the synthesis, secretion, transport, metabolism, binding, or elimination of hormones naturally carried by the blood, which are present in the body and are responsible for homeostasis, reproduction, and the developmental process.
>
> (Diamanti-Kandarakis et al., 2009: 294)

Along the same lines, they maintain that

> … from a physiological perspective, a substance that acts as an endocrine disruptor is a compound, either natural or synthetic, which, through exposures developed environmentally or inappropriately, alter the hormonal and homeostatic systems that allow the body to communicate with and respond to its environment.
>
> (Diamanti-Kandarakis et al., 2009: 294)

In this context, it is very important to note three elements that constitute firm clues for understanding the set of phenomena associated with EDs as results of social constructions: (a) they are agents external to the human body, (b) they produce interference in production and reproduction processes, and (c) they impact on the response and communication modes of the individual body in connection with its "environment."

In another framework, but with conceptual approaches that emphasize some particular features, definitions such as the following can also be found:

> A simple definition of endocrine-disrupting compounds (EDCs) has been proposed as 'chemical agent with the potential to alter hormonal action in the body.' (2.3). This implies that the exogenous agent interferes with the synthesis, secretion, transport, binding or elimination of the body's natural hormones, responsible for maintaining homeostasis, reproduction, development and/or behaviour.
>
> (Trossero et al., 2009: 60)

> Endocrine disruptor is any exogenous chemical substance with hormonal activity, with the ability to alter endocrine homeostasis by similarity, by hormonal affinity, by antagonism, by physiological interference or by modification of specific receptors, (6) that causes adverse effects on the health of the intact organism or its progeny; (8–10) its action is exerted on the fetus in utero (own fetal impact) and on the offspring of the affected case (2–4,6).
>
> (González and Alfaro Velásquez, 2005: 448)

In the context of these conceptualizations, some features emerge regarding the "scope" of EDs that have a social (and sociological) relevance of enormous importance, since these compounds affect: (1) the fetus in utero, (2) the subjects, and (3) their offspring (by epigenetic action). EDs are substances that by modifying the individual body transform the social body in the present and the future, given the multiple interactions and effects that the subjects suffer in their various interrelated positions. As a mother/father, as a fetus, as a child, and as an adult, subjects are impacted by a regulatory substance of their potency/interaction capacity (Tables 2.2 and 2.3).

Table 2.2 Some Products for Daily Use that Include Endocrine Disruptors

Deodorants and Antiperspirants	Shampoos and Conditioners	Shave Gel	Cleaning products
Toothpaste	Lotions and Sunscreens	Makeup/Cosmetics	Additives Food

Source: Prepared by the author.

Table 2.3 Some (a few) of the Most Commonplace Consequences of EDs

Neuroimmune system disorders: chronic fatigue syndrome (CFS), fibromyalgia, and multiple sclerosis (MS).	-dependent organ cancers: breast, prostate, testicular and thyroid cancer.	Damage to the female reproductive system: precocious puberty, reduced fertility, spontaneous abortions, polycystic ovary syndrome, endometriosis and uterine fibroids, premature deliveries and low birth weight, congenital damage.
Metabolic diseases: metabolic syndrome, diabetes and obesity	Damage to the neurological system	Damage to the male reproductive system: cryptorchidism, hypospadias, and reduced quality of semen.

Source: own elaboration.

The clear and forceful message is maintained by Buck and his colleagues when they emphasize in the conclusion of their work "Bisphenol A. and phthalates and endometriosis: The Endometriosis: Natural History, Diagnosis and Outcomes Study" the effects of Bisphenols:

> Continual research aimed at delineating the relationship between environmental endocrine-disrupting chemicals and gynecologic disorders such as endometriosis is paramount, and an important step for addressing larger data gaps regarding global concerns about declining female fecundity and endometriosis' association with later-onset diseases such as autoimmune disorders and cancer. We urge the continued design of novel research with innovative methodologies for investigating the relationship between environment and endometriosis at the population level.
>
> (Germaine et al., 2013: 167)

What is possible to warn is that permanent exposure to ED can alter physiological hormonal signalling (Patterson et al., 2015), that prenatal contact with these substances increases health risks of children, for example, in connection with obesity (De Cock, Legler and Van De Bor, 2011) and that it is justified to think that the presence of such and multiple substances in our bodies can be considered as catastrophes.

The epigenetic patterns of the transgenerational effects of EDs make us think about the distances and proximity of what operates as salience in these processes, and what acts as pregnancy. The modifications in time, the collective/individual edges of the latency of the effects, and the persistence of the harmfulness should make us think about the calamities that are not seen but "that-are-affecting." Although it is possible to notice the hundreds of sets of practices against the most visible EDs as emerging/rising from predation (mining, soybeans, etc.), it is also easy to notice the volume of the pregnant actions that compromise/mortgage the lifetime. It is precisely in the uneven distribution of nutrients that it is possible to see more clearly the epigenetic consequences of some of the morphogenic regularities.

2.3 Nanotechnological colonization

To better understand the impact on the human body that nanotechnology has, it is necessary to appreciate the difference that its intervention makes at the scale and structure of the cellular and molecular level. From a very elementary perspective, the cellular level comprises the smallest units of living matter. The "tissue level" implies a set of cells that perform a certain function, and it is the organ, formed by the connection of different tissues, that fulfils a function.

Every human being includes a sequence of chemical information that determines how their bodies look and how they function. This sequence is found in the long, spiral-shaped molecules, called deoxyribonucleic acid (DNA), found inside every cell. DNA carries the codes of genetic information and is made up of linked subunits called nucleotides. DNA is the chemical name of the molecule that contains the genetic information in all living things. The DNA molecule consists of two chains that wind together to form a double helix structure. Each chain has a central part made up of sugars (deoxyribose) and phosphate groups. Attached to each sugar is one of the following four bases: adenine (A), cytosine (C), guanine (G), and thymine (T). The two chains are held together by links between bases; adenine binds to thymine, and cytosine to guanine. The sequence of these bases along the chain is what encodes the instructions for making proteins and RNA molecules.

Nanotechnology is the science of manipulating matter at an atomic and molecular scale to solve problems. Nanotechnology is an applied developmental science, with the potential to make significant contributions in many fields, including engineering, computer science, and medicine. In this section (and book) reference is made to the "nano" as a scale of manipulation of the human body, to the dimensioning of biological objects or other materials implanted in the bodies and to the measures, interventions, and/or processes to which they are submitted to said bodies at different levels (cellular, molecular, atomic).

In connection with the above, it is interesting to cite here the experience of the Research Group initiated by Sara Franklin in Cambridge, under the name of the Sociology of Reproduction, as an example of a critical look at what this chapter intends to draw attention to: the identification, exploration, and intervention upon the inner planet as a territory to colonize and manipulate. On its website, the group states that it aims to develop new sociological approaches to the intersection of reproduction and technology, and that its objective is to develop more generalizable statements about the changing definitions of nature and ethics, the biologization of technology, translational biomedicine, and the political economy of reproduction. The group nurtures and contributes to sociology and anthropology, science and technology studies, social and oral history, feminist and queer theory, and the social study of biomedicine, bioscience, and biotechnology, among other fields. Among the publications of the group is that by Jent (2019) which affirms that:

> ... Inside every stem cell is an organ waiting to happen." This statement, drawn from a recent article in Nature Methods suggests that stem cells lie in waiting, suspended until something comes along to bring out the potential organ within. For postgenomic stem cell science, what enables this to happen is the stem cell's microenvironment. The bodily milieu has itself become a technology with which researchers can mimic normal and pathological organ growth in the laboratory. Paradoxically, as concerns about the Anthropocene inspire cautionary tales of human interference in the environment, manipulation of the bodily microenvironment is today viewed as the greatest promise of regenerative medicine. Stem cell researchers use notions of embodied ecologies to model development, study disease, and biomedically intervene in ageing and injured bodies.
>
> (Jent, 2019)

As it is possible to observe, the genetic, the nano, and the molecular are spaces of scientific development that must be thought in terms of a body extractivism.

In the exemplified context, let's see how the nano operates and how it allows the occupation of the inner planet. For our analysis here, it is convenient to insist that when it is said that the exploration and occupation of the body occur through external devices created at the smallest scales of life, the statement is neither metaphorical nor hypothetical. John R. Clegg and his colleagues have dedicated a review to the different forms of "bioinspired and biomimetic devices," reaching the following conclusion:

> The majority of this review progressed, in bottom-up analysis, through material-biological communication (i.e. environmental responsiveness of biomaterials), material processing and fabrication (i.e. engineering useful constructs out of environmentally responsive components) and biologically inspired devices (i.e. using materials science, micro/nano fabrication, and cellular engineering to solve medical problems). We felt that this style and the analytical sequence was logical. It progressed from fundamental principles to applied science; from molecules to biomaterials, to complex machinery; from well-known thermodynamic laws to novel work-in-progress devices.
>
> (Clegg et al., 2019: 77)

In the aforementioned study it is clearly visible how the management of nano-technology once again confronts us with the use and manipulation of EDs now implanted in the body of human beings:

> As illustrated in these brief examples, researchers are using biologically responsive polymers to construct therapeutic devices, biosensors, and bio-logical machines. Recent efforts in the field have addressed fundamental sci-entific inquiry and the development of translational technology. Fundamental advancements have included refined thermodynamic theories to explain the behaviour of biopolymers in physiological solutions; new chemistries for synthesizing diverse, environmentally responsive polymers; nano- and micro-fabrication technology for forming ordered assemblies of natural and synthetic device components; and bioprinting technology for generating cell-laden constructs.
>
> (Clegg et al., 2019: 5)

Neural interfaces play a key role in the successful intervention on one of the most delicate territories in the geography of the inner planet and it seems to be reaching increasingly surprising results:

> The seamless integration of next-generation probes with the nervous system will shed light on many unanswered questions in neuroscience, re-evaluate some principles believed to be understood, and aid in therapeutics aiming to restore proper brain function. On the other hand, future rational design and implementation of nanoscale neural devices should be guided by the physical properties of neural components and their networks. For example, to achieve 'myelinated' mesh nanoelectronic devices for modulating oligodendrocyte

behaviour, the diameter/width of the mesh filament should not be less than 0.4 μm. Efforts in optimizing these devices are still ongoing, but successes in wireless interfacing 80,136 and restoring or enhancing perceptual sensation 18,20 show that we are already at the initial stages of the nano-revolution in neuroscience.

(Ledesma et al., 2019: 14)

The world and the "being-in-the-world" are hatched from and with the regulations of sensations. The forms of the cognitive-affective schemes that constitute the aforementioned regulation emerge or "become" in a Moebesian arrangement between perceptions and emotions. Today the "modelling zones" of sensibilities are traversed by the results of genetic and nanotechnological knowledge from applied research to the valorization of capital and the central axis of ideological capital gains concerning the world made body. The point of departure/arrival of the socialization of knowledge turned knowledge of the world has its privileged inscription surface in social sensibilities.

One of the most complex facets of the social forms of the colonial situation is the character and play of the appearance and emergence of knowledge/expertise: there is nothing in genetic knowledge or in "nanos" that does not respond to a "micro-history" and a "macro-history" of its realization. There is thus a macro-history of the scientific disciplines in which they are developed, of the socially necessary technologies for their "application," and of the academic-social structures in which the funding of the two previous features has been registered. In other words, the geopolitics of knowledge cannot be analysed without the social conditions of production of the emergence, selection, production, and reproduction of said knowledge "in-the-centres" and its expansion "in-the-peripheries" as a starting point for a critique of the coloniality of the inner planet.

One of the basic characteristics of the expansion of capital is to "hide-zigzag" the connections between the production of meanings, knowledge, and knowledge spirally articulated with the valorization and commodification of life, in terms of nodes of expropriation and predation. Knowing in order to dominate and colonize the world at the "service of humanity" is part of the basic definitions that the political economy of morality provides in direct connection with the concrete practices of that regime of knowing. In this direction, the genetic and "nano" are constituent parts of the production of epistemic and symbolic violence that operates on the tensions of the horizons of possible worlds made bodies, such as "nature" and "naturalization."

If the connections of the geopolitics of "really existing" knowledge that are "ahead" of the global articulations of multinational companies are analysed, it is observed how genetic engineering and nanotechnology are supporting the value chains associated with the most profitable productive enclaves.

There are at least three of these enclaves, given their global nature, whose high impact on the countries of the Global South and strong media exposure serve as paradigmatic (and exemplary) instances of the networks that are hatched in the current context of the aforementioned geopolitics: the so-called re-commodification of dependent productive systems, automotive industries, and personal care (and home) businesses. For reasons of space and greater relevance, we mention here those that involve an articulation with the section on EDs and the nano trans-formation of bodies.

Personal and home care businesses are the third enclave in a paradigmatic sit-uation. Among the thousands of products that are produced, and the hundreds of companies that produce and market them, the volume and network of knowledge that is socially necessary for their realization is "lost." Unlike the two previous examples, this "sector" has a particularity accentuated: its high and necessary adaptation to the market. It is a set of value chains based both on well-known technologies and on the permanent development of technologies that allow it to be highly adaptable to local demand.

These three paradigmatic enclaves not only connect, at least in some of the points of the network, but are also traversed by marketing as a basic component of the expansion of capital at a global level. Although this fundamental activity of exposing the "final" ways of adding value to a product implied and involves applied knowledge, here we only want to emphasize its character as a necessary articulator in the process of the total commodification of life. A simple exam-ple is the hair fixer where the petrochemical industry intervenes, as well as the commodities and personal care businesses; since its production depends on the "waste" of one of the enclaves that are used as an input in the other, and as a final product in the third. But without marketing as a process of creation and construc-tion of its "utility" for the particular demand, it would not exist.

One of the possible indicators of these intersections and networks between productive enclaves and applications of knowledge/expertise is (are) the invest-ment(s) in research and development carried out by large companies at a global level. The sectors that receive the most investment are those that in turn increase the cross-cutting links between the depredation of common goods, technological developments, politics of sensibilities, administration of life, and colonization of the inner planet.

In this new context, the interrelationship between the capitalist economic order and political power has amplified world domination to unprecedented extremes. This domination is legitimized by invoking a new unappealable authority: sci-entific-technical knowledge and its economic, social, and individual benefits. The deep convergence of information technologies with genetics and biotech-nology, nanotechnology, and knowledge sciences, the so-called Nano-Bio-In-fo-Cogno (NBIC) has become the most forceful exponent of this authority and its enormous benefits. As the American report by Roco and Bainbridge (2003) pointed out at the beginning of the new millennium: "Converging technologies for improving human performance. Nanotechnology, biotechnology. Information technology and, Cognitive science," these convergent technologies represent the greatest revolution of all time since, among other things, as Castilla (2012) notes, it will allow individuals to expand their knowledge and communication skills, increase their physical capacities, improve their health, increase the capacities of social understanding, security and, of course, improve productivity and eco-nomic growth. In this same direction, although a little less enthusiastic about the economic and social implications, is the subsequent report of the European

Commission, "Converging technologies. Shaping the future of European societies" (Rodríguez Victoriano, 2009: 230).

One of the axes of the plots of ideological surplus value of the 21st century
is constituted by the sensibilities for "green" in terms of the "masks" of sustainable sciences. Genetics turned into accepted and acceptable social discourse, is
configured in a regime of beliefs about the manipulation of the world. The old
relationship between mass media and extra work involved in the constructions
of images of the world that provide meaning to daily life is today woven by
scientific stories about the origin, reproduction, and horizon of the survival of
the planet. A renewed way of understanding the destinies of the world as a Lay
Apocalypse where Human Being will pay for his faults or enter the kingdom of
"the chosen ones," of those included in the green paradise. Sustainability goes
hand in hand with genetics to ensure the "truly human" for "humans," through
adding value to production chains. The triad of the organic, a brand of origin and
the transgenic expresses the potentiality of the devices for regulating sensations,
as axes of a grammar of colonial action. The triangle produced by the genetic
knowledge of/about "nature" divides the inhabitants of the colonial city around
the circuits of feeding, tasting, and food; the inhabitants of the "world of no," the
good colonist, and the tourists are shipwrecked by channels marked and painted
by the genetic stories of life. Thus, fast food, slow food, signature cuisine, and
hunger coexist as colonial symptoms. The ideological surplus is served: each
style of eating is articulated with feeling practices explained in and by the genetic
story made body, literally eaten and absorbed to facilitate the classification of the
sustainable world. What transverses the ideological practices of sustainability is
the commercialization of bio-diversity as a private appropriation of a collective
good. The millions of dollars invested in the identification, manipulation, and
reproduction of hundreds of genetic maps of the environmental/energy assets,
denote one of the points through which the dialectic of the "new illustration"
passes today.

The cognitive-affective experiential structures that are produced by the inter-
relations between the nanotechnological metaphorization of life, the commercial
valorization of genetic diagrammatics, and the monopolization of the modalities
of bio-diversity are the axes through which epistemic violence passes, symbolic
and physical of colonial domination. Said structures imply a set of perceptual
frameworks that perform at least three basic "functions:" being the horizon of
understanding of "lives" in their various manifestations, constituting the "knowl-
edge-at-hand" available to the subjects to coordinate action, and socializing inadvertently the scientific analogies that involve the theories that explain the axes
mentioned above.

The nanotechnological metaphorization of life involves three moments from
and through which commodification crosses and composes the cognitive-affective:
the dissolution of what still exists between macro and micro visions of the universe, the feeling of absolute management of what exists, and the elaboration of
a scientific look at the inner planet.

The valorization of the maps and diagrams based on the genome and the
"genomatization" of the life sciences: these impact on the cognitive packages in

availability, on the logics of "appreciability" of the cognitively relevant, and on the practices of their production and "socialization."

The identification, classification, management, and production of biodiversity is one of the mechanisms with the greatest impact on the structures of feeling: it establishes a tenuous line between the diverse and the unique, it allows a breaking of the boundaries between place and storage, and it makes possible the most dissimilar ways of masking as "as if."

In the context of what has been analysed so far, it is possible to observe, at least partially, how the diagrammatics of classes (in the new colonial contexts) are intertwined with the structures of expropriation in their direct relationship with the various forms of capital gains anchored in vectorization. These inhere between the exploitation of work (in the metropolis and the colony), the dispossession of the volumes of energy inscribed in environmental assets, and in the work socially necessary to manage sensibilities. Said intertwining is better understood by observing, at least preliminarily, its connections with the narratives of knowledge/expertise of the axes of dependency exemplified in the ideological surplus value(s) formed from the genetic and nanotechnology.

Another key aspect of the application of nanotechnology is the consequences of environmental "nano-pollution" as supported by Gupta and Xie:

> To envision the health hazards coupled with engineered nanoparticles, their complete life cycle should be scrutinized from their manufacturing to storage, and from distribution to intended industrial and commercial uses/potential abuse and ultimate disposal. Furthermore, to find ways to manage and confine nanomaterials, we must continue to explore the causes and mechanisms of nanotoxicity to gain a better and deeper understanding. Ultimately, a more cautious manipulation of engineered nanomaterials as well as the development of laws and policies for safely managing all aspects of nanomaterial manufacturing, use, and recycling portends the unforeseen opportunities in this blooming field of nanotechnology.
>
> (Gupta and Xie, 2018: 228)

In this same vein and connection with the above, it is important to highlight a *Nature Nanotechnology* editorial entitled "The risks of nanomaterial risk assessment" (2020) that summarizes the problem of the consequences of nanoparticles as follows:

> Two Correspondences in this issue, one by Bengt Fadeel and Kostas Kostarelos, and one by Daniel A. Heller and colleagues highlight the problems and challenges specific to carbon nanotube risk assessment. In these pieces, nanomedicine and nanotoxicology researchers demand that carbon nanotubes should not be viewed as one type of material, as advocated by ChemSec, but that they should be grouped by their individual features, dosing and exposure routes. To achieve a fair risk assessment of carbon nanotubes, both Correspondences recommend a detailed evaluation of recent

insights into the short and long-term interactions of the different types of carbon nanotubes in the correct in vivo models and the potential adverse effects in the context of the applications for which they were designed for.

(The risks..., 2020: 163)

That is to say that they invade and pollute, occupy the objects and bodies leaving a techno footprint in their nanostructures, just as it is noted regarding cerium that this is an action of colonization that impacts constituting a new socio-biological metabolism by enrolling in problems similar to the predation of the planet in general.

One study that serves as an articulation with the next chapter, and which has already been mentioned above, is that of Pramanik and his colleagues who clearly show the connection between the IoT, nanotechnology, and the development of biosensors:

The advancement of sensor technology has paved the development of nano-scale biosensors. These nanoscale health sensors have been instrumental in modern diagnostic technologies, providing access to data from places that were previously unreachable and impossible to sense, and also from instru-ments that were previously inaccessible due to sensor size. As a result, new medical and environmental data are being able to be collected, allowing aug-mentation of existing knowledge, new findings and better medical diagnos-tics, which opens the door of the emerging medical fields and improves the overall healthcare sector.

(Pramanik et al., 2020: 65238)

Based on what has been developed so far, it is possible to observe two of the many processes that are directly affecting the state of our bodies/emotions: the unprecedented expansion of EDs and the increasingly common use of nanotech-nology. One way to "appreciate" the dimension of the impact and massification of nanotechnological developments and dissemination of EDs is to ask about the list of the ten richest people in the world, what their companies are and what they do: (1) Jeff Bezos, *Amazon*; (2) Bill Gates, *Microsoft*; (3) Bernard Arnault, *LVMH*; (4) Warren Buffett, *Berkshire Hathaway*; (5) Larry Ellison, *Oracle Cor-poration*; (6) Amancio Ortega, *Zara*; (7) Mark Zuckerberg, *Facebook*; (8) Jim Walton, *Walmart*; (9) Alice Walton, *Walmart*; and (10) Robson Walton, *Walmart*.[1] On the one hand, we obviously have those known as "retail" companies and sell-ers of goods on a planetary scale, and on the other hand, those connected to the virtual/mobile/digital world. As the society standardized in enjoyment through consumption unfolds in its fullness, you can see how enjoyment is made meat, it is clear how it socializes unnoticed the ways of being inside the inner planet: perfumes, creams, food, electronic devices, etc. All goods containing EDs, and companies that bet heavily on nanotechnology researching new materials and energies, are oriented to the same end: the unilateral accumulation of wealth based on "new coloniality."

On the other hand, if you explore the "2019 EU Industrial R&D Investment Scoreboard" made for the European Union (which is a Science for Policy publication produced by the Joint Research Center (JRC), which aims to provide evidence-based scientific support to the European policy-making process), point 6 of the executive summary finds the following statement:

> The four largest companies by R&D investment are Alphabet, Samsung Electronics, Microsoft and Volkswagen. Amazon would have been in the first place had its annual report given a figure for R&D alone so it could be included in the Scoreboard. Over the last 15 years, 8 companies have moved up in the global ranking by 70 or more places. These are Alphabet, Huawei, Apple, Facebook, Alibaba, Celgene, Gilead Sciences and Continental indicating the rising importance of ICT and biotechnology. The ranking of the top 50 large global companies by R&D intensity (all with an intensity of 13.3% or more) also highlights the importance of these two technologies with 23 companies from biopharmaceuticals and 24 from ICT.
>
> (Hernández et al., 2019: 6)

The report emphasizes the growth of biotechnology in which agriculture, animal genetics, and human health converge, clearly pointing to immunotherapy, gene therapy, and stem cell therapy as examples of the latter space.

A good indicator of the weight of biotechnology research and the expansion of the inner planet as a territory to apply it are patents. The "Scoreboard" says in this regard:

> Top 3 patenting technologies for the EU Biotech companies are 'Biotechnology' (60.5%), 'Basic materials chemistry' (10.8%), and 'Food chemistry' (7.7%), while those for US Biotech companies are 'Pharmaceuticals' (39.9%), 'Biotechnology' (26.5%) and 'Organic fine chemistry' (20.0%).
>
> (Hernández et al., 2019: 77)

The inner planet is colonized in three ways: (a) some vehicles navigate it, (b) some assets are absorbed/preyed on, and (c) there is communication with the metropolis it occupies. Cells, molecules, and atoms are explored, occupied, and appropriated by the "external" social, subjective, and natural worlds that transmit their metabolic fracture to them It is not Mars, nor any strange solar system chosen to continue predation; it is yet another means of body dispossession, now undertaken literally around the inner planet.

Note

1 https://es.wikipedia.org/wiki/Anexo:Milmillonarios_seg%C3%BAn_Forbes

References

Argemi, F., Cianni, N., and Porta, A. (2005) "Disrupción endócrina: Perspectivas ambientales y salud pública", *Acta Bioquím Clín Latinoam*, 39, (3), pp. 291–300.

Can, X. and Xiaogang, Q., (2014) "Cerium oxide nanoparticle: A remarkably versatile rare earth nanomaterial for biological applications", *NPG Asia Mater*, 6, (2014), p. e90. doi: 10.1038/am.2013.88

Clegg, J.R., Wagner, A.M., Ryon Shin, S., Hassan, S., Khademhosseini, A., and Peppas, N.A. (2019) "Modular fabrication of intelligent material-tissue interfaces for bioinspired and biomimetic devices", *Progress in Materials SciencePubMed.gov*, December; pp. 1–46. doi:10.1016/j.pmatsci.2019.100589.

Castilla, A. (2012) "Biomédica", *Instituto Nacional de Salud*, 32, (4), pp. 471–473.

De Cock, M., Legler, J., and Van De Bor, M. (2011) "Endocrine disrupting chemicals (EDCS) and childhood obesity: What do epidemiological studies tell us?", *Pediatric Research*, 70, pp. 368–368.

Diamanti-Kandarakis E. et al. (2009) "Endocrine-disrupting chemicals: An endocrine society scientific statement", *Endocrine Reviews*, 30, (4), pp. 293–342.

Estrada-Arriaga, E.B. et al. (2013) "Presencia y tratamiento de compuestos disruptores endócrinos en aguas residuales de la Ciudad de México empleando un biorreactor con membranas sumergidas", *Ingeniería Investigación y Tecnología*, XIV, (número 2), abril-junio 2013: 275–284 FI-UNAM.

Germaine, M. et al. (2013) "Bisphenol A and phthalates and endometriosis, the ENDOStudy", *Fertil Steril*, 100, (1), pp. 162–169. doi:10.1016/j.fertnstert.2013.03.026. NIH-PA

González, A.R. and Alfaro Velásquez, J.M. (2005) "Nuevos disruptores endocrinos: su importancia en la población pediátrica", *IATREIA*, 18; N4, dic., pp. 446–456.

Greally, J.M. and Jacobs, M.N. (2013) "In vitro and in vivo testing methods ofepigenomic endpoints for evaluating endocrine disruptors". *PubMed*, 30, (4), pp. 445–471. PMID: 23787785

Gupta, R. and Xie, H. (2018) "Nanoparticles in daily life: Applications, toxicity and regulations", *Journal of Environmental Pathology, Toxicology and Oncology: Official Organ of the International Society for Environmental Toxicology and Cancer*, 37, (3), pp. 209–230. https://doi.org/10.1615/JEnvironPatholToxicolOncol.2018026009.

Hernández, H. et al. (2019) "The 2019 EU industrial R&D investment scoreboard", *EUR 30002 EN; Publications Office of the European Union*, Luxembourg, 2020, ISBN 978-92-76-11261-7, doi:10.2760/04570, JRC118983.

Hoffmann, A., Zimmermann, C., and Spengler, D. (2015) "Molecular epigenetic switches in neurodevelopment in health and disease", *Frontiers in Behavioral Neuroscience*, doi:10.3389/fnbeh.2015.00120.

Jent, K. (2019) "Stem cell niches", *Fieldsights – Theorizing the Contemporary*. Published online by the Society for Cultural Anthropology (April 25).

Kortenkamp, A., Martin, O., Faust, M., Evans, R., McKinlay, R., and Orton, F. (2011) *State of the Art Assessment of Endocrine Disruptors*. Final Report.

Lerda, D. E. (2009) "Endocrine disruptors (ED) and human exposure", *Research & Reviews in BioSciences*, Regular Paper, pp. 1–6.

La Nación newspaper (2020), Monday 20 Jul 2009.

Ledesma, H.A. et al. (2019) "An atlas of nano-enabled neural interfaces Nat Nanotechnol", *Nature Nanotechnology*, 14, pp. 645–657.

Meyer, A. Sarcinelli, P.N., and Moreira, J.C. (1999) "Estarão alguns grupos populacionais brasileiros sujeitos à ação de disruptores endócrinos?", *Cadernos de Saúde Pública*, 15, (4), pp.845–850, out-dez.

Roco, M.C. and Bainbridge, W.S. (2003) *Converging technologies for improving human performance: Nanotechnology, biotechnology, information technology and cognitive science*. Netherlands: Springer. doi: 10.1007/978-94-017-0359-8

Patterson, A.R. et al. (2015) "Sustained reprogramming of the estrogen response following chronic exposure to endocrine disruptors", *Molecular Endocrinology*, Jan 16, 1237.

Pramanik, P.K.D., Solanki, A., Debnath, A., Nayyar, A., El-Sappagh, S., and Kwak, K. (2020) "Advancing modern healthcare with nanotechnology, nanobiosensors, and internet of nano things: Taxonomies, applications, architecture, and challenges," *IEEE Access*, 8, pp. 65230–65266. doi: 10.1109/ACCESS.2020.2984269.

The risks of nanomaterial risk assessment (2020) *Nature Nanotechnology* 15, pp. 163. doi:10.1038/s41565-020-0658-9

Thakur, N., Manna, P., and Das, J. (2019) "Synthesis and biomedical applications of nano-ceria a redox active nanoparticle", *Journal of Nanobiotechnology*, 17, p. 84. https://doi.org/10.1186/s12951-019-0516-9

Trossero, S.M., Trossero, N.R. Scagnetti, J., Portillo, P., and Keinsorge, E.C. (2009) "Plaguicidas en leche materna y su potencial efecto disruptor endocrino en gestantes del cordón hortícola de Santa Fe (Argentina)" (Informe Preliminar), *Revista FABICIB*, 13, pp. 59–67.

Rodríguez Victoriano, J.M. (2009) "Los usos sociales de la ciencia: Tecnologías conver-gentes y democratización del conocimiento", *Estudios Sociales*, 17, (34), pp. 226–249.

UNEP/WHO (2013) World Health Organization, United Nations Environment Programme. In Bergman, Å., Heindel, J., Jobling, S., Kidd, K., and Zoeller, R.T. (Eds.) *State of the science of endocrine disrupting chemicals – 2012. An assessment of the state of the science of endocrine disruptors prepared by a group of experts for the United Nations Environment Programme (UNEP) and WHO*. ISBN: 978 92 4 150503 1.

3 Colonization of the inner planet II
Politics of sense

3.1 Introduction

Within the framework introduced in the previous chapter, where it was possible to observe how colonization becomes a body in and through various processes, among which the diffusion of endocrine disruptors and the deployment of nanotechnology in bodies/emotions were emphasized, the need to analyse the senses appears strongly. It is important to emphasize here the connections of what was expressed in the introduction about the articulations and connections between body-image, body-skin, and body-movement and to understand in the light of Chapter 2 the body as a surface of dispute and order, of conquest and interstitials practices.

Since the beginning of the social sciences, the analysis of sensations, perceptions, and impressions has occupied a key place in the explanation of the constitution of society. For example, in Chapter 2 of Destutt de Tracy's treatise on Ideology entitled "On Sensitivities and Sensations" the author states:

> The sensitivity is this faculty, this power, this effect of our organization, or if you want this property of our being under which we receive impressions of many species, and we are aware of it.
>
> (Destutt Comte de Tracy, 1817: 28, my translation)

> These nerves in man are threads of a soft substance, about same nature as the cerebral pulp; their main trunks start from the brain, in which they meet and merge; from there, by a multitude of ramifications and subdivisions that extend to infinity, they spread to all parts of our body, or they will carry life and movement.
>
> (Destutt Comte de Tracy, 1817: 30, my translation)

In this horizon, it is clear how the senses, worldviews, and politics have been connected in classical social science as analytical elements of the social. This chapter seeks to indicate how the intervention and management of the senses is a fundamental component of the colonization of the inner planet.

3.2 Politics of the senses as a basis for managing sensibilities

The politics of the senses is the expression of the colonization of the inner planet in and through the senses that operates by three concurrent pathways: (a) by artificial mechanisms,[1] (b) by writing the biological structure, and (c) by the interface between a and b.

The largest companies worldwide devote much of their R&D budgets to the knowledge, design, and intervention of the processes/structures that generate and manage sensations. In the same vein as we suggested in the first chapter, a major puzzle of the 21st century revolves around sensibilities:

1. Brain, nutrients, endocrine disruptors, "nano," and genomic life management are the central edges of the material conditions of life in the 21st century, social ways to produce, concentrate, and play with power.
2. The social forms of the immediate entail specific surfaces where particular experiential conditions are assumed and sociabilities are supposed. The immediate is a special "here-now" that redefines the present.
3. The conditions of productivity and the commodification of sensations are united in a common "pre-origin" that is elaborated in the "use-the-body" to relate to the instruments/machines/apparatuses.

Social practices take the texture of the materiality of the body/emotions, feeling density, and processuality of the senses. Social ways of connecting with the world have moved to the hands, fingers, and the fingertips, first with communication tools, but also with the means of payment, entertainment centres, and equipment for cooking and cleaning. We live pressing keys, sliding, touching, and "clicking;" our day-to-day life is filled with cell phones, credit cards, transport payment cards, cookers, microwaves, and so on, which have been intertwined between social classes, genders, ethnicities, and ages.

In societies that paradoxically are spectacularized yet reconcentrated individually, which excite "all" but are limited to the closest, are companies that redefine the politics of touch, the conditions of the rules/rules of play, of touching, and being touched. It is precisely here that it makes sense to ask about the politics of sensibilities, as that set of cognitive-affective social practices tending to the production, management, and reproduction of horizons of action, disposition, and cognition (Scribano, 2017).

One of the central features of the political economy of morals of capital is the elaboration, "value," management, and reproduction of various and unequal modulations of the senses. The politics of sensibilities involve politics of the senses: hearing, sight, touch, taste, and smell. All politics of the senses are activities to resolve situations (*sensu* Thomas), to be successful in the social presentation of the person (*sensu* Goffman), and develop knowledge at hand (*sensu* Schütz) that subjects use, and it is in the context of these politics that touch becomes relevant.

In this way, it is possible to count a strong reconfiguration of the "politics of the senses" whose implications must be weighed in the social structuration processes on a global scale; and to value how human beings know, build, and rebuild the world in and through our bodies and emotions; through the senses. The most elemental form of the alluded contact is the social construction of the connections between sensory impressions, perceptions, and sensations. Through their bodies, social agents impact through a set of impressions in the forms of "exchange" with the socio-environmental context. Thus, objects, phenomena, processes, and other agents structure perceptions, understood as naturalized ways of organizing the set of impressions. This framework configures the sensations that agents "make" of what can be designated as the internal and external world, the social, subjective, and "natural" world, thus recreating a dialectic between impression and perception, which results in the "sense" of surplus – more here and beyond – of the sensations.

The social production of sensations implies a modulation of the senses: smell, taste, touch, hearing, and sight only to mention those that we might call "primaries." After thousands of years on the planet, human beings are the only species that can design, create, manage, and reproduce (and commodify) the aforementioned senses. The systemic intervention on the senses has generated specific politics about them; there is an expropriation and depredation of the senses. The capacities for use of the senses are distributed not only in diverse ways, but also in an unequal way.

Social relationships have been modified following a metaphor of "the press of a button," pressing "play," touching the screen, and clicking on a point to interact with the instruments and also with people. They express moving subjects, plunging sensitive parts of the apparatus, and to see or do more with this action, they have to "know how to play." This is the actual approach which, if it does not eliminate the "know-how" and "know-what" (typical of the disputes of the 20th century), redefines access to the world. The Social Bot Era is a space-time where the borderline between the superfluous, the prostheses, and the "extensions" is diluted within the "friendliness" of the interfaces to buy/enjoy. In this context, "touching" is more than a skill, it is a condition of cognitive/affective possibility of "being in the world."

Geometries of bodies and grammars of actions are organized largely by the influence of the politics of the senses. The proximity and distances between the social agents imply the social value of smell, the proper forms of touch, the acceptable ways of listening, the standards of the look, and the appetizing taste, and these are the components of a politics of the senses.

3.3 The "five" senses and the politics of sensibilities

From the point of view of current knowledge about the senses, their processes, contents, and ways of appreciating and intervening upon them, it is obvious that they cannot be reduced to five. We choose here to explore and make explicit only five given their ancestral cultural weight and central role in the structure of trade, industrialization, and financing of capitalism as we have known it in the "West" for at least five centuries.

Social studies of the senses have long been a well-established field of the social sciences and humanities, encompassing anthropology, sociology, philosophy, geography, etc. (Gabbacia, 2005, 2009; Macpherson, 2010; Kocur, 2011; Pink, 2015). Perhaps David Howes has been one of the academics who has most clearly emphasized the urgency of deepening the studies on the senses (1991, 2003). In the introduction to one of his articles he maintains:

Sensory studies involve a cultural approach to the study of the senses as well as a sensory approach to the study of culture. Such a condition challenges the monopoly that psychology has long exercised on the study of the senses and sensitive perception to highlight the sociability of sensation. History and anthropology are the foundational disciplines of this field. However, sensory studies also encompass many other disciplines and research approaches in the humanities and social sciences that have turned their attention, over the past decades, to the phenomena of sensitivity.

(Howes, 2014: 11)

It is in this framework that this section explores the consequences for the colonization of the inner planet from the classic five senses.

3.3.1 Politics of smell

In his well-known article "Sociology of Smell," Anthony Synnott provides this first overview of the "weight" of smell in society:

Each one of us at all times, emits and perceives odors, we smell and we smell, and such smells have very important roles in virtually all areas of social interaction: eating and drinking, in health, home, therapy, by reducing stress, in religion, industry, transportation, in class and ethnic relations, and personal care. Smells are everywhere and perform a wide variety of functions.

(Synnott, 2003: 431)

The smell of the different/diverse, both in its form of distinction, as a trait of nobility, and in the form of abjection as a trait of the stench of the expelled, has been and continues to be a modulation of the most important senses for acceptable and accepted body geometries. The scent is one of the edges that build mental walls and institutes circuits of racialization across the planet.

The politics of smell implies a system for the classification of subjects and environments guided by olfactory sensations that are based on multiple distinctions that establish moral values. This establishes an odour detection threshold, and with it a minimum of proximity and distance with other people and environments. The industry for body scents, fragrance for clothes, and perfumes for the home are some of the main objects of the huge business of smell.

There is a close connection between the market, smells, and societies normalized in immediate enjoyment through consumption. Denaturalizing odour by neutralizing what is bodily in olfactory interactions is an axis that runs through the coordination of actions between subjects. The smell of the other and the others inaugurating distances/proximities opens the door for a strange convergence between luxury, appearance, and pharmacopolitic of life through the perfume industry. For millennia, perfumes have been associated with gods, kings, and mythological characters, until the arrival of capitalism and the commodification of smell, a matter that has become more complex as it spans everything from body odour, through accommodation spaces, to public environments. In society, everything smells in such a way that it can be recognized, and in the 21st century that sensation is intervened upon by science, the market, and the State.

An axis that often goes unnoticed in the context of the global urbanization process is the mercantile connection between waste, garbage, and odours. From sewage treatment, through the accumulation/deposit of garbage, they have become highly complex elements in the structure and infrastructure of contemporary cities. At crossroads between public health, waste business, and the various forms of global landscaping, the "not feeling" smell becomes the objective of various public policies.

In the context of the above, smell becomes one of the fundamental nodes of the market of appreciation by which the articulations between body, skin, image, and movement can have virtuous or stigmatizing consequences for the social presentation of individuals and their possibilities of interaction.

3.3.2 Politics of taste

Deborah Lupton has conceptualized taste in a way that approximates what is expressed in this book:

> The word 'taste', when applied to food and eating, is generally used to denote the sensation people feel when they take food or drink into their mouths, linked to the arrangement and sensitivity of taste buds on the tongue. A food or beverage is described using several specific taste categories, including sweet, sour, bitter and salty, or flavours, such as lemon or vanilla, or more generally might be described as delicious, revolting, bland, rotten and so on. An alternative definition of 'taste' is the broader understanding of a sense of style or fashion related to any commodity.
>
> (Lupton, 1996: 94)

It is clear that there is in taste as sensation a great power to "model" emotions and the social relationships associated with them in the context of its special characteristic of emerging from the body/emotions.

Since time immemorial, humanity has intervened upon food and by doing so has socially constructed taste, this being one of the most committed and "used" capacities of the human body for its adjustment to the environment and is integrally connected with its reproduction. A politics of taste is based on the "between" generated by taste, food, and stress hunger/enjoyment. In today's global contemporary world, eating has become an experience of hunger for millions and enjoyment for the few. From "cuisine signature" through the mass media dedicated to the "kitchen," the academic formalization of the craft of cooking, up to the products offered as gourmet, eating and enjoying, converge in the notion of "having an experience." The geoculture and geopolitics of the various forms of commensalism, eating practices, and diets are a clear testimony that human beings link with other human beings and living beings in and through taste.

For centuries, science has intervened upon food in a double sense: to "bring it closer" to matching our tastes, and to preserve/adapt it to the rhythms of eating. Today, after a long process of modification of flavour, colour, and smell from chemistry, food modifications have had an impact on the organic and social production of flavour, on the very constitution of taste as sense. Nanotechnology has transformed not only the ingested "object," but also its relationship with the individual body and in what is socially constructed.

The colonizing force of the planetary expansion of capitalism among many converging processes has focused its attention on the displacement of the food/eating relationship towards the food/enjoyment connection.

The flip side of the food/experience is the hunger/food connection. Everything from public welfare policies, through manuals about "food for the poor," to the material conditions of access to food itself, are modifying "popular taste" crossed by the imperative of satiety. "The full belly" is a state of satiety that implies a metamorphosis in the social definition of taste and, therefore, of taste itself.

3.3.3 Politics of hearing

A politics of hearing implies the ways we configure a map of the environment, what are the components of equilibrium with the real, and a modality for classifying situations. From the recording industry, through the massive access to information, to the balance between demand/supply versus bureaucracy, there are constructed the sociabilities of listening modalities.

From the telegraph, through the telephone, to modern digital communication, human beings have colonized, manipulated, commercialized, and expanded our possibilities to transmit messages, hear the voice of others, and learn news about other places. The scribes, the storytellers, the minstrels, and the elders were not only figures of relevance for transmitting the past and identities, but also fundamentally a central part of the politics of what, how, when, and why to listen. A politics of hearing as a component of a politics of sensibilities implies a set of practices that regulate our willingness to listen. Not the person who wants to listen, but the one who can.

Hearing, listening, and auscultating are three moments of a topology of the capture of the world that begins the biological possibility of hearing, but that is transformed at the rate of the use of artificial mechanisms and ear extensions that allow sound perception in a situation of copresence, in an experience of time/space de-anchoring, and/or use of some technology as an ear extension (*sensu* Marshall McLuhan).

The politics of hearing can be better understood by analysing the flow of its impact on social relationships. In the first place, there is listening of class, gender, ethnicity, and age. Secondly, the audiovisual permeates and penetrates daily life due to its extension into the realms of habitability, education, and politics. And in third place, it generates diverse positionalities: speaker, listener, and viewer.

The importance of listening and hearing is taken up in the last chapter of this book as one of the axes for developing the social sciences of the 21st century, a centrality that lies in the fact that within the framework of the colonization of the inner planet, the autonomy of hearing and the imperative to listen become key elements of social structuration.

3.3.4 Politics of look/gaze

The proximity/distances between looking, seeing, and observing have a triple importance in today's world: (a) they are connected with the body-image as the seat of social interaction, (b) they are part of the exploration of the planet and the inner planet, and (c) are central components of the anchoring/disengaging game (*sensu* Giddens) of the copresence situation.

A politics of the look involves the connection between seeing, looking, and observing. Seeing refers to perceiving objects by the eyes through the action of light; looking is to direct the sight to an object, considering it; and observing involves the action of examining carefully. Each phase of social structuration processes has its specific scopic regime.

The sense of sight is "socially organized" in and through its preparation to identify, select, and manage others in conditions of interaction: human beings treat each other as they see each other. On the other hand, observing puts the observed action into object conditions by which society structures the conditions of acceptability, generating a market of interventions, modifications, and a cyborg body

In the context of the "societies of the spectacle"[2] (Scribano, 2017) that emerged in the last century, the predominance of the experience of the social mandate to live a life to be seen was enshrined. The connections that we will see later in Chapter 6 between the individual and the actor set people to dramaturgically perform life. A politics of sight in the 21st century involves managing the connections between the eye, the sense of sight, seeing, looking, and observing as a result of the actions of the market, state, and civil society.

3.3.5 Politics of touch

The politics of touch refers to the articulation and disarticulation between proximity/distance with other bodies, the sensory possibilities through the hand and the skin (in general) and the social values of touching. Making contact through the body with others is an object of design, construction, and management, especially in the context of the Society 4.0. In other words, at this time in the global structuration processes, there have been inaugurated renewed conditions of receptivity relationships with others (a social relationship) and the other (as a close person): the culture of touch. But appointed by a sense colonized by the "overlapping of the senses," this culture of touching and touch is a "different" way of inhabiting the world.

Phones, heaters, air conditioners, car locks, refrigerators, televisions, computers, and tablets are all or can be, at present, digital objects. At www.macstories.net/, managed by Federico Vittici, you can read about the (now already old) iPhone 7 and its haptic system:

> The haptic feedback provides tactile feedback, such as a touch that draws attention and reinforces both actions the events. While many interface elements provided by the system (for example, switches, switches and controllers) automatically provide haptic feedback, you can use generators feedback to add their feedback to views and custom controls.
>
> (www.macstories.net, 2016)

As has been suggested, we are engaged in the process of a transition to what has been called Society 4.0, and it is a social configuration strongly associated with what we can call the "era of touch." One of the threads that weaves the skein of social relationships is the one that connects the subjects in said relationships through the proximity/distances of their physical and "virtual" contacts. We live in a world organized according to the possibilities/limitations of experiencing the multiple forms of time-space unlocking that go through everyday life. Billions of people do not have "access" to the aforementioned de-anchoring, but others do so: where the conditions of copresence demanded by interactions between people have been transformed. At the same time that the telephone, Internet, and TV are modified, the very notion of playing as the central node of the rules of being with another is being transformed, and with them the possibilities of its commercialization.

We are also experiencing a Touch era that brings our capabilities from "there" to screens, device surfaces, and interfaces by simply touching, by simply sliding the fingertips over one of the aforementioned surfaces. Being and touching are mutually redefined, from the positions of the hands and fingers to their connection with sight, they are metamorphosing behind the technological changes offered by the market.

Paradoxically, the social norms around what it is forbidden to touch subsist within the framework of a state of obscenity and pornography that involves a redefinition of what is commercial in the "lifting" of the aforementioned prohibitions. The difference as monstrosity (genders, ethnic groups, ages) that becomes untouchable coexists with the sale of the "diverse" as a body-turned-merchandise that only has to be paid to be touched (assuming this in all its variations that we have been describing).

The most extreme feature of the politics of touch today is the possibility of buying a part of the body, where possession modifies the property, and with this landslide the same definition of what it means to touch another person, object, or living being.

The rugged paths of playing that crossed and/or skirted multiple limits from autonomy/heteronomy/recognition "valued" are transformed into mere characteristics of the property of who touches and what is touched.

The colonization of the inner planet in terms of meaning is complex and of great impact; conjugating a plexus of inquiries that go beyond one of the senses and seek results in their interactions to organize and commodify them.

Given the specific anatomical overlaps between olfactory and emotion-related neural structures, olfaction stands out in the sensory landscape for its particular and close relationship with the world of emotions. Odours affect behaviour, mood, and well-being, as well as cognitive processes such as memory and preference acquisition. The importance of perfumery through the ages attests to the close links between odours and phenomena such as emotions (Baer et al., 2018).

The colonization of the connection between smell and taste is clearly seen in the inquiries to manage the reception and consumption of tastes and fragrances, as evidenced by the study by Donato Cereghetti and his colleagues. In this study they present LikeWant, an innovative behavioural research method that measures the motivation of consumers to seek flavours and fragrances as a reward; and where they showed that the LikeWant procedure is capable of (1) measuring the desire for a pleasant odour with odourless air as a neutral control condition, and (2) discriminating between two fine fragrances based on their rewarding properties, potentially allowing the use of the procedure in consumer studies.

In a similar vein, Mars Roel Lesur and his colleagues in their article "Being short, sweet, and sour: congruent visual-olfactory stimulation enhances illusory embodiment" look for the most appropriate modality to associate smell and sight, following the trans-modal and multimodal journey to manage perception. Although they consider the results preliminary given the experimental setting, they point to associations between odour and objects that potentially play a role during impressions of illusory property. They think this may be relevant to better understanding of the multimodal nature of the body sense, as smell has been an often-forgotten modality.

In the field of connections between the visual and taste, the work of Alexandra Difeliceantonio and her colleagues is very clear in its conclusion regarding the relationship between incentives and food:

> These results imply that a potentiated reward signal generated by foods high in both fat and carbohydrate may be one mechanism by which a food environment rife with processed foods high in fat and carbohydrate leads to overeating.
>
> (Difeliceantonio et al., 2018: 10)

The study that was carried out clearly shows the connection between reward and food selection and its impact on consumer's decision-making.

In a more radical vein, Lehner and his colleagues in their article "Monetary, Food, And Social Rewards Induce Similar Pavlovian-To-Instrumental Transfer Effects" indicate:

> Our study was designed to investigate the influence of different reward types on behaviour when each reward type was calibrated to the same subjective value. A key feature of our study was the modified (Becker DeGroot-Marschak) BDM using motor effort, instead of money, as a common currency. This allowed us to successfully match all the different types of reward to the same subjective value and then, conduct a well-controlled PIT experiment.

Even though the monetary and food reward were not matched directly to each other, but rather calibrated independently to a specific subjective value, the inferred exchange rate between chocolate and money reflects the actual market value of a packet of Maltesers® surprisingly well (0.15 Swiss Francs per piece of chocolat)"

(Lehner et al., 2017: 8)

3.4 Knowing touch and politics of the senses

From our point of view, the central axis for reflecting on the "digital relationships" is to understand the changes from the standpoint of operating surfaces. One of the important points is the vehicles for registration on these surfaces: the eyes, fingers, and ears. There is thus a triangle of expropriation of surplus instantiated in watching, playing, and listening, as these senses are reconfigured in our societies. Consider, to contextualize this metamorphosis, the implication of the expansion of "touch" as technological mediation through the daily lives of millions of people from relatively "short time" ago.

The politics of the senses plays a triple role in contemporary societies: (a) it impacts on the geometry of bodies and the grammar of actions, (b) it is a nodal component of the rules and norms that shape the institutions of the State, the market, and civil society, and (c) is part of the central "architecture" of what in Chapter 6 is called the "dialectic of the person."

Societies normalized in immediate enjoyment through consumption imply the distance that operates from a politics of touch where the other cannot be touched and must be avoided because they are the object of a brand that infects and threatens. Distance is defined by appearance and rostricity, the other becomes an object interpreted by what he is wearing, by what he appears to be, and by what he is socially said to be. The other is condemned by his face, he is valued because he seems dangerous. In society you learn to value by looking at the face, you learn just by looking that the other can be a threat, so you learn to label, to name, and to make a prejudiced gaze performative.

Contemporary society is spatialized in cities, towns, villages, and landscapes that are thought from their limits, assumed from their edges, and inhabited about their margins, as expressions of sensibilities that read their disposition of "place" as a result of socially elaborated eyes, noses, mouths, hands, and ears as compasses and as storage containers for social history made body. The "bodies-in-contact" make up the schematicities of the webs of the senses turned into emphatic mediators of contacts, touches, and pacts. The politics of the senses involves the geopolitics and geoculture of feeling as the spatialization of that special combination between past, present, future, and presentification.

Being with another through the sound, smell, image, taste and folds of the body can unfold as a condition of possibility of the social relations that are the traces of inhabiting the world. If we concentrate (and stop) on the relationship between the senses and the city, then the marks, lines, and thicknesses are followed. If one looks at the landscapes, settings, and contexts which limits, borders, and margins draw and colour, a series of connections can be captured between various dispositions of the city at their intersections with the politics of bodies and sensibilities. Some of these are the following: a) the policies of the entrance, accumulation, and circulation of food structure and order the dispositions of the forms of the city; b) the policies of origin, circulation, and use of water are preconditions of survival of the inhabitants in terms of the axes of nutrients, classes, and body geometries in the form of differentiated habitability; the policies of the social forms of management of sunlight distribute, attribute, and impute the capacities of order, displacement, and visibility of the agents-made-bodies; c) the structures of selection, management, and accumulation of waste imply policies of administration of odours and redefinition of the "climate of possible predation;" d) the multiple "routes" of transfer, visualization, and contact affect and are "pre-designed" according to class, age, ethnicity, and gender; and e) the distribution structures of energy sources to eat, shelter, move, and communicate are the knots of a network woven around its accesses and possession.

How is it possible to perceive the social forms of coexistence that populate the times/spaces that are woven primarily in and through the senses? The tensions between the multiple experiences of eating, the hours and diagrams of natural and artificial light that are accessed, the differential and unequal distributions of water, the modes of movement, and the points of contact between those who inhabit the city are based on and are a fundamental part of the politic of bodies and current emotions.

The "social organs" are the ones that guide, locate, and accompany the displacement through and in space, place, and geography; it is the sensation regulation devices that are "used" as a reservoir of possible orientations to these social organs. The lives of people serve as existential ways of presenting, in and through the body, conflicts and affective encounters in the diversity of spatialized sensibilities.

Forms of predation, especially those linked to the expropriation of bodily and social energies, take shape in urban networks and locate/dislocate the subjects involved in them. The experience of the spatialization of the senses provides time-space to the surplus expropriations that performatively make the expropriator and expropriated see, smell, touch, and hear each other through the politics of sensations.

In this context, some of the vectors with more "weight" appear in the consequences of mimetic consumption for the structures of sensibilities, namely: the politics of the gaze, the politics of the mimetic image, and the politics of instant enjoyment. Precisely these three politics are configured around the obliteration and avoidance of what is conflictual in the colonizing life forms. The Mobesian band from which they start, bifurcate, and twist is an unnoticed, naturalizing, and inadvertent situation of conflict between the expellers and expelled, segregating and segregated, and colonizers and colonized. The multiple intermediate "locations," intervening locations, and blurred positions that are constructed in the folding and refolding processes of the three aforementioned policies, are those that guarantee, testify to, and endorse the opposing tensions that generate them.

The politics(s) of the gaze(s) is(are) designed in the torsion produced by being willing and disposed for classificatory observation, the seduction of what is seen (and reseen), and the labels of rostrofication[3]. The one and the other, the fears, and the aggressions are made (in principle) from the "distance" of the look that narrates, produces, and/or eliminates the other from the vision, perception, and imputation of threatening/friendly, own/improper, and close/foreign traits, impacting on the (im)possible socialities. That is, it is about the provisions of classification and taxonomy of what is and what can be expected, and how far the observer can approach without care, without defences, and without barriers. Along with this, a practice of production of the individual "to-be-seen" is carried out, as the axis of a permanent state of seducing for acceptance, recognition, or at least the "non-rejection" that is based on appearing as it is, needing to get around the "scanner" of multiple classifications. This is where the face emerges with force as the "place" of the gaze that consecrates difference, inequality, differentiation, and distance. Some labelled faces, typical faces, and "ethical" faces are identified, selected, ordered, consumed, and "circulated" as axes of the political economy of current morality. It is in this context that in our cities a glance can cost a life, allow access to enjoyment, or make travel (im)possible.

From the base of the second twist described in the politics of gazes (the individual "to-be-seen"), the possible bands of the politics(s) of the mimetic image(s) are opened(s). In its webs of limits, edges, and margins, in the city scenes, landscapes and contexts of life are constituted to be shown/hidden. The dialectic between body, skin, image, and movement is resolved and strengthened as "practices-for-others." In the same tension, the experience of the city appears as multiple shop windows, stained glass windows, and catwalks directly related to consumption. The mimetic image is a consequence of the differences and inequalities in the consumption of bodies and emotions as objects, as a source of labile, contingent, and context-bound "identities." Every city that "boasts" has routes of detection, classification, selection, and imitation of possible images. These are processes that will make viable and liveable a valuable and valued habitability according to the material conditions of the selection of each one of them.

In the meanderings of the connections between the politics of the gaze and of the mimetic images, that is, in what is in them in common with consumption, strongly spatialized politics of instant enjoyment are raised and designed, and it is precisely this instant enjoyment that underlines the range of temporality acquired by a set of practices "dependent" on spaces, and especially on which limits, edges, and margins are performed.

Instantaneous enjoyment is characterized by being a "here-now" parenthesis, thanks to its claim to be continuous in time, and by the subjective disengagement that it allows and demands. The enjoyment is made as a circumstantial, contingent, fleeting, but "absolute" and radical "here-now." Enjoyment is an act with the pretense of totality that suspends the flow of everyday life, hence it is "made," produced, and performed, and is the result of a teleological action. It is enjoyable to continue enjoying it. In reality, everything in life is taken and consumed with (and according to) enjoyment. This form of enjoyment detaches the subject from its context: the one who consumes/enjoys is not "that-subject," he is only an enjoyer. The subjective unlocking is done through the primacy of the object of enjoyment that varies according to the quantity/quality, volume/density, and forms of access/denial to it.

These frameworks between the politics of the gaze, the politics of the mimetic image, and the politics of instantaneous enjoyment make visible the distances and proximities between the marks of the edges of the autonomy/heteronomy game, the lines of the limits of the dialectic between expulsion/destitution and the thickness of the margins.

What is, is placed and is placed in (and through) the limits, edges, and margins are practices of feeling that stabilize, fluidize, coagulate, and mobilize the "logics of occupation" of the city. Be these objects, subjects, or relationships, all are subject to the operation of sensibilities that are produced in the devices for regulating sensations and social bearability mechanisms that nest in these forms of being that the city has. For this reason, (and in this) the emergence of a colonizing city is verified by occupying, expropriating, making it possible to inhabit the time-space of another, and by containing the power to decide on the lives of others, expands its effects, and consequences "beyond." The world in general, but particularly in the Global South, has viewed the performative consequences of the limits, edges, and margins that make the politics of emotions cut to the waist of the politics of the gaze, of the mimetic image, and instant enjoyment. The differences and inequalities in the capacities to design, apply, and live the politics of sensibilities lie in the measurements, densities, and volumes that the material conditions of existence grant to the marks, lines, and thicknesses that the city creates.

It is in this framework that the politics of the senses are an object of reflection, design, elaboration, and management as direct ways of colonizing the inner planet: eyes, hands, skin, nose, mouth, and hearing are the landing beaches built to make the body in the image of, and in similarity to, the social.

3.5 The visual experience of the world and inner planet

The perception that is located closer to "the look" that constructs the "visual datum" must be inscribed in specific regimes of sensibilities challenging the empiricist feature that is imbued with the "datum" as a reality that exists independently of the eye that looks. Thus, a space for critical reflexion opens up to deal with the notions associated with the "representation of what is real" that usually runs through common sense; due to this, it is necessary to explore the possible connections between perception and the units of sense that are our object of inquiry.

In this point, it is important to underline the place of the chromatic metaphor of social relationships like a way to connect structures and processes with the colonization of inner planet and the politics of senses.

Thus, establishing some analytical keys about these regimes contributes to enriching the process of making theoretical and methodological decisions for research in/from the visual on/through the Net.

Based on the idea that the subjects' as well as the researcher's "looks" are inscribed in enclosed regimes of sensibilities, we propose to shift from the notion of "visual datum" to the conception of what Scribano has proposed about the *units of experiencing*. In fact, challenging the notions of units of observation and analysis, the author sketches a concept that is adequate to clarify the processes of observation, recording, and analysis of social sensibilities, emphasizing the urgency of redirecting the researcher's perception to a *hiatus* that opens between analysis and observation (Scribano, 2011).

In his proposal, focusing on sensibilities involves understanding that this critique of perception (in whose theoretical background we can recognize the expected objectives and results) is oriented to registering: (a) the space of inter-action linked to showing (us) and showing (oneself); (b) the complexity of the dramaturgical situation (Goffman); (c) how, from where, whom, and what the registered expressions tell; (d) the capacity to register "silences" (absences); and (e) the expressive weave between the experienced sensations and the emotions (Scribano, 2011: 23). If the *units of experiencing* are perceived in the process by which "...what the subjects *feel*, what the subjects do to manifest what they *feel*, and what is *felt* by the subjects who receive/look/share what is made, are assembled (disassembled)..." (Scribano, 2013: 81), then the *visual experiences* constitute an effort to limit the said process to what has historically been defined as the "sense of sight." The specificity of "the visual" mediated by the Internet abandons the ascetic appearance of the datum, hence emphasizing that its particu-lar "texture" becomes relevant precisely because it implies the porous relation-ship between perception and sensation. It is on the basis of this complexity that an experience of/with "the visual" on the Internet could strengthen "what the image communicates," inquire about "what is absent from the image," and emphasize "what the image manifests" (Scribano, 2008).

What we call "political economy" of "the look 4.0" is related to a global phase of humankind in which our senses converge (in a way that is different from how they had up to this point) with the world of technology in the search for "assis-tance/artificial mechanisms aids" to say what is real and how we experience it. The old pre-eminence of the eye – and the sense of sight – in our modern society is restructured about the complexities that we have reconstructed in this chap-ter. It is in this context that it becomes relevant to highlight the content of the practices that are changing our ways of producing sociabilities, experiences, and sensibilities to be able to map possible roads for our future research practices.

The first issue that we would like to emphasize is related to the conditions for receptivity of images in our societies. These days, from telephones to heaters, all sorts of "devices" surrounding us are or can be digital. We are immersed in a process that is called revolution 4.0, and that is strongly associated with what we call *touch culture*.

The social forms of immediacy entail specific spaces in which conditions of particular experiences are taken for granted and sociabilities are presupposed. What is immediate is a special kind of "here-now" that redefines the present.

Conditions of productivity and commodification of the sensations converge in a common "pre-origin" that gets elaborated into "using-the-body" to relate to instruments/machines/devices. In this context, social practices acquire the texture of the materiality of the body/emotions, the density of sensation, and the natural processing of the senses. The social ways of connecting oneself with the world have been transferred to the hands, fingers, and fingertips: firstly, with our communication instruments, but also with payment systems, entertainment systems, and cooking and cleaning devices, etc. We spend our lives pressing buttons, sliding, touching, and "clicking."

Social relations have shifted towards a metaphor of "touching," "clicking," and interacting with instruments. Subjects express themselves by sliding, pressing sensitive parts of devices and thus, more than being able to see or do, it is necessary to "be able to touch." This is an approximation to what is real which redefines the accesses to the world even though it does not eliminate "the know-how" and the "know-what." We are in a post-cyborg era in which the line between the superfluous, the prosthetic, and the extension gets diluted with the "friendliness" of the interfaces to buy/enjoy. "Being able to touch," more than a skill, is a condition of cognitive/affective possibility of being in the world. The 21st century will be a century of "touching," and the social sciences will have to redefine themselves in terms of their inquiry strategies and ontological discussions. In these contexts, actor, agent, subject, and the author will be reconfigured and consequently, their public positioning will be modified as well.

Image and "the logics of looking" are not immune to these transformations; on the contrary, they are framed in these redefinitions of what *seeing-sensing-touching* implies. This is *the second issue* that we would like to highlight.

The digital image that we have referred to in this chapter emerged in a context of technological and productive revulsions, of institutional sliding, and of the resurgence of the importance of the hand as the organ that is connected to thinking. Seeing-feeling-oneself starts with a touch-looking-at-oneself. Today (more than ever?) the image is an intersubjective production that acquires features of an instantiated practice at the moment of acquiring the production made for those who see themselves. While we produce an image, we touch the surfaces of devices that we need to look at to see-ourselves-feeling what we want to know and let others know.

Today, to see is to touch, feeling what is seen. Fingertips come into contact with the screen(s), the glass is pressed when a decision is made, and when the fingertips slide, they surf the menu of options that the previous selection made available. When we see a photograph, we touch it, at times in an almost imperceptible way, but most of the times with that moment of monitoring that stops incorrectness: the unwanted *like* the incorrect *upload*, the wrong *stalking*.

In this manner of interaction, human beings are elaborating a grammar of vision as a code that is closer to us than the word. The digital image on the web is a proposal to have an experience that involves getting immersed in a scene that goes through the sensibility and sense of publishing images with our hands.

The production of the digital image on/through the web is guided by capturing, not photographing, looking for a capture – not a photo – trying to convey an experience – not an object – in a massive and radically self-produced way. It is a synthesis of a scopic regime which, based on what is old, produces "new" consequences. Even though all images try to convey experiences, images on/ through the Internet are based on the quality of "portraying" something and they use that as a starting point.

Are the transformations in our scopic regime transformations in our value system? If all aesthetic transformations correspond to certain ethics and, in turn, certain politics, then the answer may be that they are. Accepting that we are "feeling-thinking" beings (in the sense of Fals Borda) leads us to wonder about our "video-touching" condition as producers of sensibilities that enable us to know/ sense the world. The challenge for the social sciences of societies standardized in immediate enjoyment through consumption in the context of the revolution 4.0 still is how to connect/disconnect science and politics.

In the context of Chapters 2 and 3, there emerges a question about how the social structures that operate as scenarios of the colonization of the inner planet appear, and that is dealt with in the next chapter that seeks to provide an approach to the mechanisms and their constitutive features.

Notes

1 With artificial mechanisms, an allusion is made, partially, to the discussion that for We live pressing keys, sliding, touching, and "clicking;" our day-to-day life for several years has already taken place under the horizon of the cyborg body. CFR Buran, S. (2015), Haraway, D. (2006), Henne, K. (2020) Preciado, B. (2013).
2 In another place and in the footsteps of Debord, Bataille, and Baudrillard, I describe the basic contents of the sensibilities that are lived in spectacle; and the paradoxical status of the sacrificial.
3 Action of classifying human beings through the face.

References

Baer, T. et al. (2018) "'Dior, J'adore': The role of contextual information of luxury on emotional responses to perfumes", *Food Quality and Preference*, 69, pp. 36–43. doi: 10.1016/j.foodqual.2017.12.003

Buran, S. (2015) "Correspondence between Cyborg Body and Cyber Self", *Journal of Research in Gender Studies*, 5, (2), pp. 290–322.

Destutt Comte de Tracy, M. (1817) *Elements D'Idéologie. Première Partie. Idéologie Proprement Dite*. Paris: MVe Courcier, Imprimeur-Libraire.

Difeliceantonio, A.G., et al. (2018) "Supra-additive effects of combining fat and carbohydrate on food reward", *Cell Metabolism*, 28, (1). doi:10.1016/j.cmet.2018.05.018

Gabbacia, D.R. (Ed.) (2005) *Empire of the Senses: The Sensual Culture Reader*. Oxford: Berg.

Gabbacia, D.R. (Ed.) (2009) *The Sixth Sense Reader*. Oxford: Berg.

Haraway, D. (2006) *A Cyborg Manifesto: Science, Technology, and Socialist-Feminism in the Late 20th Century*. Netherlands: Springer.

Henne, K. (2020) "Possibilities of feminist technoscience studies of sport: Beyond cyborg bodies." In Sterling J. and McDonald M. (Eds.) *Sports, Society, and Technology*. Singapore: Palgrave Macmillan. doi:10.1007/978-981-32-9127-0_7

Howes, D. (Ed.) (1991) *The Varieties of Sensory Experience*. Toronto: University of Toronto Press.

Howes, D. (2003) *Sensual Relations: Engaging Thesenses in Culture and Social Theory*. Ann Arbor: University of Michigan Press.

Howes, D. (2014) "El creciente campo de los Estudios Sensoriales", *Revista Latinoamericana de Estudios sobre Cuerpos, Emociones y Sociedad-RELACES*, N°15. Año 6. Agosto - noviembre 2014. Córdoba. ISSN: 1852.8759. pp. 10–26. Available at: http://www.relaces.com.ar/index.php/relaces/article/view/330

Kocur, Z. (2011) *Global Visual Cultures: An Anthology*. Oxford: Wiley-Blackwell.

Lehner, R., Balsters, J.H., Herger, A., Hare, T.A., and Wenderoth, N. (2017) "Monetary, food, and social rewards induce similar pavlovian-to-instrumental transfer effects", *Frontiers in Behavioral Neuroscience*, 10, pp. 247. doi:10.3389/fnbeh.2016.00247

Lupton, D. (1996) *Food, the Body and the Self*. London: SAGE.

Macpherson, F. (2010) *The Senses: Classic and Con-Temporary Philosophical Perspectives*. Oxford: Oxford University Press.

Pink, S. (2015) *Doing Sensory Ethnography*. London: SAGE.

Preciado, B. (2013) *Testo Junkie: Sex, Drugs, and Biopolitics in the Pharmacopornographic Era*. New York, NY: Feminist Press at the City University of New York

Synnott, A. (2003) "Sociología del olor", *Revista Mexicana de Sociología*, año 65, núm. 2, abril-junio.

Scribano, A. (2008) *El Proceso de Investigación Social Cualitativo*. Buenos Aires: Prometeo.

Scribano, A. (2011) "Vigotsky, Bhaskar y Thom: Huellas para la comprensión (y fundamentación) de las Unidades de Experienciación", *Revista Latinoamericana de Metodología de la Investigación Social- ReLMIS*, 1, (1), pp. 21–35.

Scribano, A. (2013) *Encuentros Creativos Expresivos: Una Metodología Para Estudiar Sensibilidades*. Buenos Aires: ESEditora.

Scribano, A. (2017) *Normalization, Enjoyment and Bodies/Emotions: Argentine Sensibilities*. New York, USA: Nova Science Publishers.

4 Structures/relationships/processes

4.1 Introduction

This chapter must be understood in the context of the observations set out in the Introduction and Chapter 1 of this book, where we argue in favour of understanding the colonization of the inner planet in terms of the co-connection between "politics of the bodies," "politics of emotions," and "politics of the senses;" this emphasizes that they are political because their modulation and execution must be thought in terms of the politics of sensibilities that are elaborated in the tensions between sociabilities, experiences, and sensibilities. These politics manage sensations, emotions, and design sensibilities, they are constructed as a set of feeling practices that configure the political economy of morality. Sociability, experience, and sensibilities are thus the basic structures of the colonization of the inner planet. It should be noted here that in a complex way all these processes acquire the shape of a spiral, network, and/or Moebius tape.

In this context it is necessary to make clear that there is no possibility of understanding processes and structures without analysing the practices of language and speech acts. Social practices – from the characteristic Weberian definition as intentional action, through Mead's linguistic mediation, to Habermas's communicative action, Giddens's recursive practices, or Bourdieu's practical sense – are based on the acceptance of the place of meaning and language in its constitution and reproduction. Language, in all the functions that can be attributed to it, is an inseparable element of meaning as an expression of experiences and perceptions. In and through language we human beings live our lives and reproduce them. In the social sciences, and especially in the programmes that are of interest here, the "linguistic turn" had an impact in such a way that it recontextualized the discussions around the subject, the intersubjective relationships, and "the frames of meaning."

In the context of the above, and within the framework of the book in which this chapter is inscribed, it is also necessary to underline that there is a set of analytical perspectives on the connection between language and emotions; schemes that in one way or another incorporate, are a part of, or discuss the "weight" of what we could call the landscape of emotions.

A similar (though slightly different) example, akin to the approach that we support here, is that proposed by Wilce:

> We gain much, as well, by examining not only local contexts of the relationship between language and emotion (from the micro-interactional to the 'cultural') but also the globally social context of this relationship. There are political economies of language and feeling, and their variously envisioned connections. Cultural concepts of feeling and language are partial, interested (not disinterested) representations, often linked with the political economy and thus can be usefully described as ideologies.
>
> (Wilce, 2009: 13)

From a dissimilar perspective, it is possible to verify that at present different analytical approaches combine emotions and language. One of these is the analysis of feelings, which can also be performed in different ways: (a) one of them is related to the use of computational tools and the creation of semantic dictionaries, "[...](t)he method of sentiment computing based on the semantic dictionary is mainly rely on the open-source sentiment dictionary or the extended sentiment dictionary, and the sentimental value can be calculated combine with some semantic rules" (Zhang et al., 2018: 396); and (b) there are also multimodal visions that emerge from the same social complexity,

> [...] [a] number of recent studies have attempted to recognize sentiment expressed in social multimedia from multimodal signals, including visual, audio and textual information. [...] Typically, speech transcripts along with facial and vocal expressions are analysed separately and the results of unimodal, text-based sentiment analysis are fused in post to form a 'multimodal sentiment analysis' system.
>
> (Soleymani et al., 2017: 4)

For more than a decade we have been aiming to account for the importance of the "existential turn" in social theory (Scribano, 1998), advocating a close connection between the studies of the body and emotions (Scribano, 2007a, 2007b, 2011; Luna & Scribano 2007), and also supporting the importance of exploring a line of study regarding the intersection of these works, by investigating the place and feeling of colours about the issues that they raise. In all these works, how sociability and experientiality have a direct connection with the overflow of language through expressiveness and the creation of a plurality of languages have been made explicit in various ways.

The social structuration processes at the planetary level imply a fundamental redefinition between State, Market, and Civil Society in such a way that the accepted and acceptable politics of sensibilities are modified and with them the structures, relations, and social processes. Within this frame, it is necessary to observe that connections and relations between the management of bodies and the marketization of emotions have turned into a central axis of the current social structuration, and thus constitute one of the basic challenges for social sciences in our century.

In this book, it is possible to see how "the political" is intertwined with "the emotional." The globalization of emotionalization serves as the central axis of the current metamorphosis of relations between state and capitalism, between politics and market, and between "ideology" and marketing.

4.2 Mechanisms, processes, and experiences

An explanation of the social that accepts (and starts from) the intersections and frameworks produced by critical hermeneutics, the critical theory of the Frankfurt school, and dialectical critical realism must be able to account, among other questions, for the ways to analyse and articulate mechanisms, processes, and social experiences. In what follows we propose the intersection of the epistemological and theoretical "substantive" spheres to carry out the aforementioned analysis.

4.2.1 Being for the fruit or the germinal logic of the mechanisms

From the power and place of the metaphor as a component of the structure of a theory, we refer here to the relations between seed and fruit as a way of explaining the constitution of social mechanisms. The dialectic between departure and arrival, between origin and destination, and between phylogenesis and ontogenesis, is able to be expressed metaphorically through geminal logic.[2] From here, new practices of knowing are envisioned (and proposed) that neither run aground in causal and objectifying reason, nor do they become entangled in the paralysis of a negative dialectic.

The logic of the seed starts from the assumptions of the implication between the phenomena and the actions of people. There is no doing that does not self-implicate the doer in the implications of the doings of others. There is no doing (self) that does not include that density and porosity of the collective that, by reinventing itself, transforms, and transforms individuals in the dialectical continuity of social relations. To observe the pathways of the mechanisms "*that make things happen*" it is necessary to jump the wall from the mere causality of the fetish to the germination of the fruit.

In said inscription surface, the observer ceases to be someone with knowledge and becomes a "being-in-knowledge;" it ceases to be a machine for immediate possession of effects, and becomes a participant in an uncertain possibility of sharing flourishes in the context of the uncertainty of the complex. Fruits are not only born, they grow, develop, and die; also, and fundamentally, they are processes in production and production in the process. In other words, when doing producer constitutes the world, that same world configures that doing as a process that makes a world another. The fruits follow the dialectical path of *appearing*, *leaving their trace*, and *reappearing* in various forms. The mechanisms understood from the logic of the seed are always presented as multiple, flexible, and intertwined with their own history.

The fruit appears, is presented as an affirmation, as the *first moment* of reality as a mechanism of "locating" through "where things happen." The fruit is a positivity that depends on the history of its seed, which in it acquires meaning insofar as it appears as the content of the vicissitudes of its contingency. Here the expression "things happen" is taken in two directions: as space where the action takes place and as a generating device.

In this sense, the germinal is configured from the logic of the footprint, the *second moment* of understanding the devices that, "in passing", allow us to make evident the connections of what appears and appeared with the new forms that they take in other conditions of time-space those first affirmations. The trace that leads from the seed to the fruit allows the rebuilding of an effective functioning of the mechanisms through signs.

As a *third moment*, the germinal strategy is presented in the structure of the fruit, in its conditions of appearance, in its constitutive elements, and in that "rare" quality of being willing, on the one hand, to subsume appearance and trace, and to be prepared for suppression in your ingestion, for one another. There is in the seed–fruit relationship a tensional quality between deconstruction and structure; the dialectic of a "be-being" for the fruit is evident.

These three moments – appearing, leaving traces, and structuring – that are inferred from the dialectic of the germinal, allow us to "put together" an explanation "by-means-of" the mechanisms that make the social instantiate, take place, and dissipate in its metamorphosis.

In this same direction, but from a transversal look at the production of germinal structuring mechanisms as a mode of explanation, it is important to pay attention to the edges, folds, and shapes that this has, thus complementing the possibility of understanding the colonial situation. One dimension originates by covarying the germinal in the dialectic of the real made in the multiple determinations of the concrete. The edges can be understood as double shapes of the qualitatively unfolded bands; links and double loops as the logic of a dialectic of action through the germinal. The superposition of the edges reconstitutes the space-time fabric of the mechanisms, interrelates them under the modality of structuring, and shows the covariance of the transformations between said mechanisms: the order-change vision comes out of its binary moulds. The constitution of the fruit housed in the decomposition of the seed points in the direction of a triple action of the mechanisms (and their interactions): components, acting factors, and decomposition. The unpredictability of the multiplicity of acting factors is "limited" to the processes required to carry out the indeterminate step of composing and decomposing. The metaphor of the germinal recovers the micro-history of the mechanisms, a "story" that appears at the edges of their transformations as the tension between precipitating elements and expected results.

The dependent neocolonial situation can be explained from a dialectic that, accepting the complexity and uncertainty of the metamorphosis of the universe of capital, is on track for a germinal logic as an alternative to a geopolitically centred causality.

4.2.2 Moebius tape, spiral, and network

The tension between mechanisms and processes, their continuity and discontinuity, their mutual influence, and their dependence on the event-structure flow leads us to recover a series of concurrent paths for analysis. In this section we propose the *Moebius tape*, the *spiral*, and the *network* as forms that complement the analysis of the dialectic of domination; as figures that express in various ways what is precisely dialectical in it.

In the context of metaphorical and sociological use of qualitative geometry and Escher's designs, the exposition of social processes in terms of Moebesian bands is intended to allow the visualization of the moments of the run-through, of folding and unfolding, of transversal gaze that are needed in order not to duplicate the real in the mere representation of a binary space. Social processes open–close like the bands of a Moebius tape that when cutting them multiplies into another band. From this perspective, this social "geometric space:" (a) transforms the visualizations of the proximities and distances between phenomena; (b) updates a look to the bias avoiding the specularity of a linear look; and (c) allows identification of the blockages of the processes by qualifying the torsions produced in them.

In this way, the importance of a look at the bias appears in terms of following the torsions produced by the proximities and distances between the phenomena. The specular version of colonial reality is disproved as soon as the obturations of a mirror that does not reflect, but is assembled in the process of looking, become evident.

What is in the social processes of flow and repetition, of inherent and contingent, of succession and rupture, is possible to be evidenced through the form of the spiral. The spiral unites the circle, the arrow and the point as ways of spatializing time. They pass and go through the same place following in the direction of another from the moment. Understanding processes as spiralling actions of *ascent-descent-overcoming makes it* possible to visualize what is dialectical in them. In this sense it: (1) is possible to identify the moments of production, reproduction, and transformation of the phenomena inscribed in a flow of events; (2) hints at the influences, reflexivities, and recursivities between the phenomena in the context of their occurrence; and (3) implies the potential to recognize in an element of a "higher" phase the presence of another element of a "lower" phase, and vice versa. A spiral vision of social processes makes it possible to capture the flow of contradictions, oppositions and transformative manifestations existing between the phenomena developed in them.

Social processes and the phenomena inscribed in them can be better understood if the analytical figure of the network is applied to them. Exchanges, interactions, and flows; the connections, derivations, and directionality between nodes of a reticular structure allow accounting for the multipolar characteristics of the processes at different spatio-temporal scales. In the aforementioned direction, the network form: (a) imputes meaning to the visible and invisible, blocked and exposed, and poured and inverted connections of the interactions between phenomena; and (b) allows us to identify, select, and classify the nodes through which the consequences are crossed-tied-unleashed.

It is in this framework that the relations between imperialism, dependency, and colony can be thought through the figure of the spiral, the Moebius band, the global networks configuring the dialectic of the domination of colonial situations. In this context, it is possible to understand imperialism without a single crown, dependency without a single metropolis, and colony without a single occupying army as features of the current state of subjection relations.

These three ways of presenting subjection on a planetary scale are indeterminate, complex, and changing forms that the capitalist system of exploitation and expropriation adopts to shape, maintain, and reproduce itself. These forms of capitalist subjection are currently hatched and re-hatched as "openings-closings" and "entrances-saliences" of the folding and unfolding of a Moebius band. These forms distance themselves, approach, and remain equidistant; these forms appear, cancel each other out, and reappear in a dialectic of domination that deepens, sharpens, and overcomes their relational character. The development and withdrawal of the multiple determinations of the concrete are constitutively tied to the historicization of the connections and disconnections that occur in the (dis) articulation of the contingent of the imperial, dependent, and colonial situation of the Global South today.

4.2.3 The chromatic metaphor of social relationships

Everyday life, the flow of our sharing, and living the world is experienced from, through, and with our body. In the morning, upon awakening, the predominant sensation is that of inhabiting a territory, experiencing a space, and carrying a structure. In this way, our body appears to us as the closest and most unshakable and, at the same time, as the strangest and most malleable. Many times during the day, our body is presented to us (and we narrate it) from a chromatic analogy. We feel "radiant" or "opaque," "lucid" or "dull." Through our body we experience life painted in different colours: we have black or pink days. The state of social and bodily energies are the first manifestations of social conflict. The social world of possible lives that subjects can experience is the result of multiple intersections between the social and bodily energies that we have, the antagonisms[3] that we star in to produce and reproduce these energies, and the social modes that are developed to narrate and understand them.

Our lives are the result of spellings made bodies thanks to the obstacles and potentialities housed in the differential distribution of corporal and social energy. Our lives are the result of practices of autonomy and subjection that can be mapped and painted according to the tones that we are capable (or not) of printing upon them. One way that can be used to describe these body games is to construct a conceptual guide that uses some elements of colour theory. That is, moving to a chromatic metaphor that allows an understanding of the connections and disconnections between sociability, experience, and sensibilities, and can be used like a conceptual critical instrument to examine the politics of senses, as shown in Chapter 3.

Understand social actions from the founding metaphor that implies the reality that colours involve: the game between energy, bodies, and the observer. It is no accident that theories about colours were originally built from pictorial "experience." It is precisely a picture of the social world that can be observed if the distributions of corporal and social energy are taken as a starting point, at least partially. It is also possible to visualize from this metaphor the modes of social sensibility that appear from said distribution.

Just as there are no "essential" colours, but rather different ways of reflecting the light that bodies have with a given observer, in society body tones are built based on the body and social energy that agents can manage in ranges of autonomy differentiated according to their class, positions and conditions, and the bodily tones of the one who does the looking. Synthetically it can be affirmed from this perspective that: there is colour in bodies, there is colour in how bodies look at each other, and there is colour in how agents look at (experience) the painting of the world in which they are inscribed. Through these experiences, social practices can be described by the tonality, luminance, saturation, achromatism, and colour blindness that they imply.

The forms of capitalist domination are manifested through bodily marks, one of those marks consists of the ability of those bodies to reflect the state of their bodily energy. Social relationships are symbolic and chromatically mediated interactions, and due to this last feature, bodies are presented and represented with differential energy intensity. In other words, in the work of approximation, the definition of situations and orientation of social action, the agents "carry" and put into play their bodily energies that are understood and "used" with and by the other agents. The quantities, qualities, volume, and intensities of energy available to agents involve the differential capacity that these bodies have to paint the world.

The logics of domination as a set of superimposed and tensional articulations imply the expropriation, predation, and subsumption of social and corporal energies. From this point of view, life in society can be described and criticized by a chromatic interpretation scheme. In other words, the material conditions of life operate as spaces for the expansion of "wave frequencies" where bodies receive and absorb certain objective "possibilities" of visibility-invisibility, associated with the social energies that these material conditions determine. These objective possibilities are reflected and/or perceived in the form of a chromatically differential body structure, where inequalities in the capacities to absorb and/or repel the determination of the aforementioned material life conditions have repercussions on the ways of representing and representing themselves that these subjects and classes of subjects have. These capacities and these ways of representing themselves constitute schemes of perception and appreciation that are translated into social processes of chromatic classification of bodies.

Some of the proposals have been synthesized in the analysis of the structural: the germinal logic for the analysis of the mechanisms, the figures of the Moebius strip, the spiral, and the network as analytical paths of the dialectic of the processes and a chromatic theory for capturing experiences. These proposals are a possible (but privileged) way to restore the elaboration of theories in and from the Global South, the place of the network between structures, relationships, and experiences cancelled by the colonial postmodernities that are accomplices of the global expansion of capital and the colonization of the inner planet. Thus, we understand that it is necessary to reweave and re-string the connections between structure(s), process(es), experience(s), and emancipation(s), assuming the contradictions of the social sciences as a science ready for happiness and not for excess expropriation; this is the question that Chapter 7 takes up in terms of hope.

In the next section, the appearance of the "subsiadano," the assisted or subsidized citizen, is analysed because it is a mix between subsidized and citizen, breaking dialectically with the citizen and producer/worker dichotomy.

4.3 The assisted citizen: the "subsiadano"

The colonization of the inner planet is part of a Moebius band for which a special form of citizen appeared in the 21st century, attesting to the close connection between consumption, state subsidy, and citizenship. The subsiadano is an example of the structural consequences of a process characterized by a serial and opaque chromaticity generated by an unequal accumulation of body energy.

In the last 30 years there have been two major structural transformations: (a) the redefinition of the connection between public policies, social policies, and politics of sensibilities; and (b) changes at work. Due to the constraints of space, it is impossible to expose them systematically here. In a synthetic way, and as a "framework" of what is developed in this section, it is possible to synthesize them as follows.

Since the end of the last century the connections between state bureaucracies, public policies, and social policies have redefined the "new social question." State, market, and civil society have developed a set of strategies to reduce conflict and manage exploitation. From the increase in the conditions of poverty and unemployment in the global order, capitalist states today have decided to intervene in the so-called "vulnerable" population, beyond the type of country and forms of government, essentially through conditional cash transfers, as the ways by which a planetary process of colonization of the sensibilities can be established. This is demonstrated by the fact that these types of devices are applied from the United States, continuing through Europe, and reaching Latin America, Asia, and Africa, as a way to improve and guarantee the reproduction over time of the poor assisted sectors and generate peaceful population management (De Sena, 2014, 2016, 2018).

In the context of the increase in informality, unemployment, and the feminization of the workforce, one of the modifications of work, of work management, and its connections with the politics of sensibilities, focuses on the emergence of conflict networks, social activities led by emerging actors (call-centre workers, social networs workers, platform delivery workers, among others). It is in this sense that it is possible to verify how a series of mechanisms aimed at extracting people's vitality (bodily and social energies) operate, as well as how emotionalities associated with certain occlusions and dislocations of collective actions are configured. Here also appear the processes by which the "new" and dynamic work scenarios linked to the digitization of life affect the possible connections between sensibilities and social conflicts. Thus, the relevance and expansion of "digital work" as a significant component of the daily experience of workers can be entered as a context of what is examined in this section (Scribano and Lisdero, 2019; Lisdero, 2012, 2017, 2019).

The objective of this section is to make evident the emergence of a "new" position of person and citizen in the context of the society normalized in immediate enjoyment through consumption: the assisted citizen. One of the main consequences of the connections between social policies and the politics of sensibilities is the "creation" of a modality of subjectivity based on the close relations between consumption, assistance, and enjoyment.

Over the last 20 years, at least, democracy by consumption has elaborated the stage for neoliberalism as a politics of sensibilities. Maybe the best example is that of the World Bank (1997) statement about "changes on State roles" that announces the "new" relationships between market, state, and social policies (Chhibber, 1997).

The last phase of neoliberalism begins with the advent of "progressive governments" and radical acceptance of capitalism as an immutable social system.

Beyond all kinds of discussion (Scribano et al., 2018, 2019), a concrete and "hard" element of the progressive governments was the acceptance of their reformist character. In the inner logic of this acceptance lives the deepest consecration of reality.

The market needs state aid to warrant the capacity of people to consume, and by this ensure the all-dominant economic system. One way to characterize this phase is to take into account the economic role of immediate enjoyment through consumption: the mixture and sum of mimetic and compensatory consumption have resulted in a state that warrants market profit.

It is in this way that the components of a political economy of morality that allows the reproduction of the capitalist system today are the consequence of the birth and death of neoliberalism as a political regime. Likewise, the alluded components make possible the consecration of enjoyment as the centre of life and the forgetting that it is possible to change the social world. Anxiety, non-movement, freedom without autonomy, and a daily life ordered around compensatory and mimetic consumption, is the end of neoliberal history.

In this context it is clear: neoliberalism is part of the populist cycle and flow, but in a deeper sense we need to see some of the characteristics of democracy by consumption, especially the assisted exploitation and subsidized consumer.

Progressive governments of the last decade configure a new way to "embody" the components of neoliberalism: economic growth is delivered through consumption policies, related to these are applied for conditional cash transfers, and a big part of Gross Domestic Product comes from the international price of commodities. The phantoms of poverty are disbanded or diluted through the fantasy of compensatory consumption. As in Foucault's analysis, social policies became a central key for managing class conflict. A paradox has been installed: the government that comes from social movements dissolves collective action into the public and private spectacle of enjoyment.

The convergence between the public policies oriented to ensuring a corporate profit and the expansion of the market to commodify the sensations have created the main opportunity for the reproduction of neoliberalism's style of capitalism. The consumers there aren't citizens, at least in the "classical" way. They are enjoyment seekers without political motivation. More or less two decades of progressive government have resulted in the insulation of people in consumption.

The tension between citizen and consumer takes shape as the "subsiadano," a key part of the challenge to understand the "political" for the social sciences in the 21st century.

The creation of the citizen took place in France and the "West" throughout the long period from the French Revolution to the Second World War. The introduction of the practices/narratives of rights was incorporated into the political economy of the already-established moral positionalities of producer/consumer and customer/citizen whom thinkers from Parsons to Habermas intuited as the axes of the post-war State. Rightly called the successive crises of the welfare state are the provisions of consumer, producer, and citizen which undergo a profound transformation. The worker/producer from Fordism, through Toyotism, reaching to Uberization and digital work has been characterized by disappearance, casualization, and permanent transformation.

The consumer has existed for more of half of "20th-century contemporary history:" in the form of the "comfort seeker," or like the "one-dimensional man" (*sensu* Marcuse), or as a serial addict to fashions, or in its latest form as one dependent upon immediate enjoyment. There is a direct passage of the citizen–voter of governments to the quasi-universal elector via consumption and "acceptable" sensibility. In this way, state practice implies a transversal orientation to gender, ethnicity, and class actions that continues with support programmes directed towards compensation policies. These tensions between producer/consumer/citizen are evidence of the basic "organizational" features of a state that is undergoing a profound transformation.

By 2015 Latin America had more than 135,000,000 people receiving conditional cash transfers (De Sena, 2016). To this figure must be added the millions of individuals who are "owners" of other programmes and living in precarious conditions of "assisted liability" by state intervention or omission. Also, if we include the numbers of citizens who receive subsidized transport, energy, and/or basic services (just to mention three activities), the millions of subjects increase and multiply, and show a relentlessly structured trans-classed subsidy. Suffice it to recall only the declared intention of the World Bank to show an increase in the middle classes. Finally, it is important to incorporate in the analysis the propensity for sustained high participation in public employment in the formal labour force which in some places has become the only source of income.

If we consider the current conditions of state aid and state action strategies, we are not only citizens in the Global South, but we are subsidized. In this context, it becomes relevant for the democracies to include citizens who are "content" (using a medical analogy) and "happy," that is, as suggested by the etymology of the word. From this point of view, it is state practices that establish the resignation (it is not possible to effect any transformation) as a consequence of the logic of patience and waiting (you must wait for your turn) as "civic virtues."

Relations between subsidy and citizenship are established through:

1) A systemic consecration of rights as untying narrativeswith the real individual (*sensu* Marx). Both in the "axes" of the central countries and the Global South there has operated an increase of the "consecration of rights" without effective guarantee of their update, and it is possible to observe how the "generations of rights" occur, encompassing increasingly aspects once considered to be the realm of the private or something protected or regulated by the State: consider the right to enjoy implants and robotics and/or informational interfaces. And in its (painful) obverse we see hundreds of thousands of

victims of multiple wars for whom the aforementioned "implants/interfaces" become necessary complementary forms of humanity. More dimensions of life with codified rights do not imply more resources or dignity in reality.

2) Redefining the connections between "lack of," consumption, and suture. The subsiadano votes as he buys, buys how he feels and feels like he is assisted The State is no longer the only actor responsible for public policies, now it is the market that must ensure the subsidized spaces of everyday life. The market as an agent that elaborates the condition of possibility of consumption must guarantee the possible sensibilities of the subsiadano.

The expansion of the state actions is proceeding through the market: it is agreements with private capital that determine subsidized areas as hubs for the long-term reproduction of capital. The State "does consume" to make the company responsible for the producer/consumer synthesis, thus creating the condition of possibility of a rational subsiadano: an "opportunity seeker" defined by their cunning to achieve consumption at lower prices and to improve the purchasing power of their income.

The subsiadano is a fundamental component of the financialization of daily life: everything in instalments, all on credit. It is the mass of the American middle class white human who lives his life in instalments and who "can lose everything." Capitalism has retraced his steps by restructuring the relationships between savings/consumption/sacrifice/luxury/credit. The 21st century is the dialectical tension between these practices through a political economy of morality whose backbones are sensibilities.

Collecting, receiving, and "use" are the political practices of subsiadano: collection by a public policy of aid, receiving conditional cash transfers and use of the benefits of agreements between the State and the market. Participation in cooperatives, micro-enterprises, canteens, among other practices, make subjects collect a reward monthly (and with a credit card) that installs a systemic instability between employment and work. Updatable subjects receive amounts of money with which they must take action to ensure the continuity of the aforementioned reception: bring children to medical checks, make them attend school, etc. The subsiadano "enjoy" subsidies given by the State to companies supplying gas, water, energy, public transport, etc. Behind this enjoyment, there is a systematic increase in corporate profits disguised as an aid to citizens.

Assisted exploitation is the superior phase of flex-exploitation and depressed desire. The compensatory and mimetic consumption works as a vehicle of deep and inadvertent sensibility construction. The dispossession of a capacity to make connections between desires, pleasure, and enjoyment is the pillar of normalized society.

The perfect milestone to democracy by consumption and assisted consumption is the exploitation through aid. The new function of social policy is to make possible the extraction and refocussing of bodily energy. These two processes are made a reality by immediate enjoyment through consumption: the people lose desire and pleasure and replace them with instantaneous and indeterminate enjoyment.

The logic of global capitalism implied the shift of colonization from the external/ "natural" world to the inner planet: exploitation, expropriation, and dispossession are the three faces of neoliberalism as a politics of the sensibilities.

4.4 Internationalization of the emotionalization regime

The globalization, local massification, and global spectacularization of the commodification of the elaboration, management, distribution, and reproduction of emotions have become the competitive and motivational core of capitalism. Feeling, experiencing, having an experience, and connecting with objects/subjects, are the most demanded and produced goods in current capitalism. The difference of this stage/moment of the dialectical structuring of capitalist restructuring is that these emotions are not qualities of an object. On the contrary, they are the objects that are requested, acquired, consumed, and discarded. These commercialization imply, of course, their solidary obverse that, in turn, are redirected back to the entire production system: the various forms of solidarity, the different forms of social responsibility and, especially, business responsibility, are there to testify how sensible one can be to the marginalized of the compensatory consumption of sensations, and the "value" that this acquires as the axis of a reproduction "on a human scale" of capitalism. This scenario is articulated/complemented by policies of state compensatory consumption and virtualization of life that combine and involve its financialization and correlates with the entertainment industry.

Inscribed in the aforementioned articulation, it is possible to notice renewed connections between the internal market, the international conjunction of elites and dependent middle classes, and the internationalization of the emotionalization regime. In the Global South, in Latin America and Argentina, for example, in the last 40 years, there has been a strong transformation in the notion and scope of what is called the internal market, given the intense processes of co-construction between the "international 'and the' local." There is not only a solid interdependence concerning what is produced, its components, and the "primary" sources of goods, but also in their distribution and marketing. What and how much to produce is less and less a local decision.

Politics, the State, and the market intersect and emerge in and through emotionalization: understanding for this the consecration of "the emotional" as a criterion of validity and verification of the political economy of morality. One edge that crosses the tensions between internationalization of the elites, emotionalization, and the market is the virtualization of life, as an emphasis on its consequences for financialization/banking and the entertainment industry.

Today in life nothing can be boring, everything must be displayed, and merchandise must be available anywhere in the world. Through this simple route, the political economy, the mass media, the political economy of morality, and the political economy of public works are connected "again."

Within the framework of technological gaps, scientific dependencies, and the coloniality of formats, today there is the internationalization of content (Netflix, news, and sports channels) that has become an "overcoming" of the already old purchase of "canned programs." In fact, globalization, multinational cell phone,

Internet, and entertainment companies, the effective banking of billions of human beings, and the massification of electronic means of payment (such as cards and PayPal) are contextual features that enhance "access" to the world-in-fun. Today it is obvious to maintain the centrality of entertainment in the structuring of everyday life: sporting events, musicals, and artistic manifestations of various kinds complete the programming of TV, radio, and topics of interest on social networks. However, what has been a "novelty" for several years is that such entertainment is organized emphatically around emotions.

Entertaining implies, at least, three knotted practices: (a) it is a separator of experiences, (b) it is a distraction turned into a goal; and (c) it is a method of suturing:

a)   "Between-having (oneself)" is to operate a separation in the flow of experience, it is to elaborate a "between" in contiguity of actions and to put a parenthesis in the fluidity of an iterative life.
b)   It is also "getting lost," as an objective in itself. Stop carrying the burden of a task and set said omission as a goal and do not find the original path.
c)   It involves at the same time filling in the gaps, hiding the faults and, above all, spending time.

Now, these three characteristics of entertaining are tied to the reality that said practices are designed, elaborated, managed, and reproduced from the global commodification processes associated with entertainment corporations, States, civil society, and "social business responsibility policies" led by multinationals. Thus, the old formula of entertainment is abandoned as a repairman for the tedium of work to consolidate the paradigm of the entertaining reproduction of life. In this framework, in the context of internationalization of emotionalization, so that life continues to reproduce enjoyment as the opposite of boring, it is necessary to appeal to a set of compensatory consumption policies.

4.5 Structure of compensatory consumption policies

Today, state intervention in society is focused on the elaboration of policies that compensate for capital/labour inequalities, faults in the identity constitution processes, and the gaps in enjoyment produced by the conditions and class positions. The moralizing guidelines of the aforementioned policies are elaborated, managed, and reproduced in and through the emotions.

Social policies, by creating sociability, also build experiences and sensibilities. They do so in such a way that what is shared inadvertently by management practices with the assumptions of the theories, become a body. The social made body is knotted and woven with the embodied statehood, thus including in the lives of the subjects a certain experience from the results of the dialectic between state practice and social practices.

In close connection with the above, and as a metonymic expression of the phenomenon, a strong link is established: the practices of statehood are related to the practices of a society normalized in immediate enjoyment through consumption.

The explicit intention of the economic policies of the current democracies ("progressive" or "neoliberal") in Latin America is to seek growth by increasing domestic consumption, where its massification plays a role of fundamental importance.

Economic policies are articulated in a "virtuous" way with a set of social policies, especially with Conditional Income Transfer Programs, in such a way that in the last decade millions of Latin Americans have been incorporated into consumption via state assistance. Consumption styles, growth of the consumer classes, compensatory transfers, and elimination of tensions, are clear expressions of how state administrations place in the expansion of consumption the key role of avoiding conflicts, re-functionalizing the participation of millions of subjects in the market, and redefining citizens as consumers. It is in this framework that we have argued that compensatory consumption is the solidary obverse of conflictual decline and resignation. But what does it mean to compensate?

Compensating is the attitude of restoring part or all of property lost, stolen, and/or dispossessed by some people to others. It involves the plot of actions aimed at repairing the injuries, offences, and/or grievances of some subjects against others. It involves an action for compensation for damages, losses, and/or torments caused. Damages, injuries, and/or losses caused are not only due to reflexively monitored actions, but also to the unintended consequences of the action. Losses, grievances, and/or damages should not be considered only as intentional actions in individual subjective terms, but also (and mainly) as those arising from class conditions and positions, unequal structure and differential access to goods, and/or occupied places in the systemic processes of dispossession, expropriation, and excess appropriation.

Compensating is a private practice that became a State practice that was once used for conflict avoidance and assurance (in time) of the capitalist "rate of profit." Compensation, understood in this way, was the key to the Keynesian welfare state and the source of its ability to stabilize the capital/labour conflict. The set of public policies that were called neoliberal were aimed at dissolving state mediation, redefining the conditions of compensation, and orienting state practices towards the goal of ensuring reproduction via privatization.

Compensatory consumption is a process that falls within the folds of current accumulation regimes, state compensation systems, and the expansion of market logic. Compensatory consumption is today the main public policy aimed at reinstalling the effectiveness of "modernity" as the cement of colonial societies. Far from being characterized as neo-developmentalists, in the Global South today the forms of State must be thought as and from their accumulation regime. Transversally, "adolescent capitalism" (as opposed to its supposed senility), has structured a set of political regimes that makes the expansion of consumption its main policy aimed at stabilization and conflict elision. The metamorphosis of the State has produced sociabilities, experiences, and sensibilities that, like the great companies and world corporations, are designed to the extent of the production, management, and reproduction of sensations. The classic passkey of the State as a mediator of the conflict, which consisted in the production of wage goods (education, health, tourism, etc.), has shifted to its ability to generate a "type" of consumption that fulfils triple functions: (a) it naturalizes predation, (b) it expands the capacity for the reproduction of the various fractions of the capitalist classes in power, and (c) it grants the necessary means for the consecration of immediate enjoyment as the axis of daily life.

The emphasis on policies of expansion and consolidation of consumption, both as a "redistributive" mechanism and as devices to expand internal markets and production, have placed consumption at the centre of the scene of the practices of coordination of action between subjects, between classes, and between subjects and the market. The network of market-subjects-State relations has been re-interlaced in, by and through consumption, presenting the consumption-production-wage-consumption circle, with the "virtues" of the "good-for-all," producing three basic consequences (with multiple reproduction bands of each one of them): (a) re-establishing the social fantasy of the social connection via the market, (b) endowing upon consumption the magical power as the beginning/end of well-being, and (c) re-individualizing society in terms of immediate enjoyment through mimetic consumption.

From the above, it is possible to notice what consumption compensates for:

1) It compensates for the faults/failures of the unequal distribution of immediate enjoyment.
2) It compensates for the distances between social fantasies, as far as there are devices for regulating sensations, and the material conditions of consumption.
3) It compensates for possible connections/disconnections between mimetic consumption, enjoyment, and coordination of action.

Societies oriented towards immediate enjoyment structured around mimetic consumption and intervened from compensatory consumption tend to discourage processes of social protest and reproduce a politics of sensibilities that transits between indifference and resignation. For the compensatory emotionalization/consumption dialectic to become "concrete" it is essential to specify specific features of the political economy of morality in and through Public Infrastructure (PI). These material conditions are the result of a joint action between the State and the Market of various magnitudes, intentional coordination, and scope: (a) a "glocal" system of travel as enjoyment consecrating tourism as a fundamental social and subjective practice; (b) a glocal system of knowledge, circulation, and reproduction of food as a specific and special experience, making "gourmet culture" a parameter for modulating action; and (c) a glocal system of spectacle based on the "rock concert" format that makes possible the production and sharing a specific collective experience and magical flavours.

Practices, all of them anchored in very specific times-spaces, in which electricity, water, organization of access, and basic facilities are provided or guaranteed by the State, and where the activity is offered by the Market, often subsidized or "contracted" for the state. These practices, all of them, are based on a common node made up of three sides: experience, sensibility, and "ineffability." Travelling, eating, and attending the recital become one of the axes of economic policies around the senses that every State that thinks about its reproduction must ensure.

One more tension of the Moebesian tape involved in the approximations/ distances between PI, public policies, and social policies is the international character that the first involves (in any of its forms of infrastructures), implying a consolidated process of participation by multinational companies. Dependency is a global phenomenon, today with redefined and multicentric actors.

Shows, entertainment, infrastructure, and consumption are four sides of a geometry that outlines the conditions for the possibility of colonizing the inner planet.

4.6 Mechanism and new dependencies

As it is possible to notice, the diagnoses that were offered in Chapter 1 converge in a process of restructuring the dependency mechanisms that imply a direct connection with the material conditions of existence to guarantee the colonization of the inner planet.

Returning to the above, it is possible to note that:

a) The globalization, local massification, and global spectacularization of the commercialization of the elaboration, management, distribution, and reproduction of emotions have become the competitive and motivational nucleus of capitalism.

b) In the context of internationalization of emotionalization, so that life continues to reproduce enjoyment as the opposite of boring, it is necessary to appeal to a set of compensatory consumption policies.

c) Today, state intervention in society is focused on the elaboration of policies that compensate for capital/labour inequalities, faults in the processes of identity constitution, and gaps in enjoyment produced by class conditions and positions. The moralizing guidelines of the aforementioned policies are elaborated, managed, and reproduced in and through the emotions.

d) For the dialectic of emotionalization/compensatory consumption to become "concrete," it is essential to materialize specific features of the political economy of morality in and through PI.

e) PIs have become the cement of societies, not only because of their character as organizers of life, but also because they stand as the glue of morality of enjoyment.

Faced with this situation, the possible dependencies in the context of new forms of imperialism and colonialism become the object of urgent redefinitions. In this horizon, it is possible to better understand what has been argued at the beginning of this work: it is possible to characterize the modalities of dependency(s) (social and individual) from and as the loss of autonomy. The volumes/spaces/margins of autonomy of subjects and groups have been cut/confined in and through the public/state management of emotions, sensations, and perceptions.

The internationalization of emotions becomes a truth regime based on the demonstration of the sensitization effect. This effect is defined by the general acceptance that what makes practice interesting, valuable, and worthy of being reproduced is its ability to move sensations and build a state of affection. In this way, what these processes manage to produce becomes felt and shared truth. The subjects appreciated are the "sensibilities" ones, the valued foods are those that "make you have an experience," the aesthetic and cognitive criteria are fun and enjoyment in specific combinations for each case: clothing, partner selection, work, etc.

On the other hand, PIs are designed, built, and managed to enhance and facilitate immediate enjoyment through consumption, the conditions of transfer and reproduction of the workforce, and the consolidation of institutional regimes of politics. An underground connection between highways and streets in cities to facilitate access to workplaces and administrative centres by improving the "social mood;" some squares, parks, and a multipurpose centre for recitals, exhibitions, and/or fairs (food, clothing, technology) for mass enjoyment; and airports, bus terminals, trains, boats, and planes for the repairer and well-deserved vacation break. All PI is felt, wanted and "necessary." A "work" is evaluated both for its contribution to well-being and for its impact on feeling well. The subsidized persons transfer their analysis of "opportunity cost," firstly, to weigh which part is free/subsidized, secondly, in how it contributes to the enjoyment and, thirdly, how it impacts on the reproduction of their material conditions of a lifetime. In this way, the political economy of morality subsumes emotionalization, public work and consumption, elaborating emotional standards to value life. Paraphrasing Marcuse, we intend to emphasize our interest in enjoyment practices because they have become political categories.

In its Moebesian tension, the internationalization of the emotionalization regime, the structures of compensatory consumption policies, and the political economy of the morality of PI open the doors to new dependencies; a scenario that challenges us to continue investigating these practices empirically.

In the next chapter, a scheme for understanding a geometry of the person is elaborated, by which it is intended to provide a conceptual frame to capture the position of the human being in this new cosmos that is the inner planet.

Notes

1 The expression in Spanish that refers to the union of subsidized and citizen as a new modulation of the relationship of people with the State is maintained here.
2 Without a doubt, in this section, the echoes of Rodolfo Kusch are evident beyond the clear differences with his ideas. https://es.wikipedia.org/wiki/Rodolfo_Kusch
3 The connections between conflict, antagonisms, and disruptions are discussed in Chapter 6.

References

Chhibber, A. (1997) "The state in a changing world", *Finance & Development*, September 1997. Available at: https://www.imf.org/external/pubs/ft/fandd/1997/09/pdf/chhibber.pdf
De Sena, A. (editora) (2018) "La Intervención Social en el inicio del siglo XXI: Transferencias Condicionadas en el Orden Global". Buenos Aires, ESEditora. 290 páginas. ISBN 978-987-3713-26-2. Disponible en http://estudiosociologicos.org/portal/la-intervencion-social-en-el-inicio-del-siglo-xxi-transferencias-condicionadas-en-el-orden-global/

De Sena, A. (2014) *Las Políticas Hechas Cuerpo y lo Social Devenido Emoción: Lecturas Sociológicas de las Políticas Sociales*. Buenos Aires: Estudios Sociológicos Editora/ Universitas. Editorial Científica Universitaria. ISBN 978-987-28861-9-6. Available at: http://estudiosociologicos.org/portal/lecturas-sociologicas-de-las-politicas/

De Sena, A. (2016) *Del Ingreso Universal a las "transferencias condiciona-das", itinerarios sinuosos*. Buenos Aires: Estudios Sociológicos Editora. ISBN 978-987-3713-13-2. Available at: http://estudiosociologicos.org/-descargas/ese-ditora/del-ingreso-universal-a-las-transferencias-condicionadas/del-ingreso-univer-sal-a-las-transferencias-condicionadas.pdf

Lisdero, P. (2012) "La guerra silenciosa en el mundo de los Calls Centers", *Papeles del CEIC*, N° 80, ISSN: 1695-6494, Universidad del País Vasco, Marzo de 2012, España. Pp. 1–31 Available at: http://www.identidadcolectiva.es/pdf/80.pdf

Lisdero, P. (2017) "Desde las nubes... Sistematización de una estrategia teórico metodológica visual", *Revista Latinoamericana de Metodología de la Investigación Social – ReLMIS*, N°13. Año 7. Abril - Septiembre 2017. Argentina. Estudios Sociológicos Editora. ISSN 1853-6190. Pp. 69–90. Available at: http://www.relmis. com.ar/ojs/index.php/relmis/article/view/213

Lisdero, P. (2019) "Labour, body and social conflict: The digital smile and emotional work in Call Centres". In Scribano, A. and Lisdero, P. *Digital Labour, Society and the Politics of Sensibilities*. NY: Palgrave–Macmillan. ISBN: 978-3-030-12305-5.

Luna, R. and Scribano, A. (Comp.) (2007) *Contigo Aprendí... Estudios Sociales de las Emociones*. Córdoba: Universidad Nacional de Córdoba.

Scribano, A. (1998) "Complex society & social theory", *Social Science Information*, 37, 3, pp. 493–532.

Scribano, A. (Comp.) (2007a) *Mapeando Interiores. Cuerpo, Conflicto y Sensaciones*. Córdoba: Jorge Sarmiento Editor.

Scribano, A. (2007b) "¡Vete tristeza... viene con pereza y no me deja pensar! Hacia una sociología del Sentimiento de Impotencia." In: Luna, R. and Scribano, A. (Comp.) *Contigo Aprendí... Estudios Sociales de las Emociones*. Córdoba: Universidad Nacional de Córdoba.

Scribano, A. (2011) "Sociology and epistemology in studies on social movements in South America", *Sociology. Thought and Action*, 28, 1, pp. 131–148.

Scribano, A., Korstanje, M. E. and Timmermann Lopez, F., (Eds.) (2019) *Populism and Poscolonilism*. UK: Routledge.

Scribano, A., Lisdero, P. (2019) *Digital Labour, Society and the Politics of Sensibilities*. NY: Palgrave–Macmillan. ISBN: 978-3-030-12305-5

Scribano, A., Timmermann Lopez, F., Korstanje, M. E. (Eds.) (2018) *Neoliberalism in Multi-Disciplinary Perspective*. Palgrave Macmillan. ISBN: 978-3-319-77601-9.

Soleymani, M., Garcia, D., Jou, B., Schuller, B. Chang, S.F. and Pantic, M. (2017) "A survey of multimodal sentiment analysis", *Image and Vision Computing*, 65, pp. 3–14.

Wilce, J. (2009) *Language and Emotion. (Studies in the Social and Cultural Foundations of Language)*. Cambridge: Cambridge University Press.

Zhang, S., Wei, Z., Wang, Y. & Liao, T. (2018) "Sentiment analysis of Chinese micro-blog text based on extended sentiment dictionary", *Future Generation Computer System*, 81. https://doi.org/10.1109/ACCESS.2019.2907772

5 Notion of the person

5.1 Introduction

In all colonization, there is a colonized that serves as the axis of the appropriation and dispossession of the different forms of energies, and especially of the bodily energies. It is a colonized domain that implies and makes evident the instantiation of the occupation made future and its epigenetic success. As Mbembe argues,

> The second principle is linked to the territorialization of the sovereign State, that is, to the determination of borders in the context of a new imposed global order. The ius publicum quickly takes the form of a distinction between, on the one hand, those regions of the planet open to colonial appropriation and, on the other, Europe itself (where the ius publicum must perpetuate dominance). This distinction is, (...), decisive when it comes to evaluating the effectiveness of the colony as a formation of terror. Under ius publicum, a legitimate war is largely a war conducted by one state against another or, more precisely, a war between 'civilized' states.
>
> (Mbembe, 2011: 38)

In the colonization of the inner planet where the transfer of the metabolic fracture to the same body/emotion is taking place as the "last" territory of conquest; the dispute and the reconstruction of those designated as individual/actor/agent/subject/author is one of the axes of the social structuration processes.

Individual/actor/agent/subject/author are at the same time modalities of designating the dispositions of the person, and therefore its geometry and its contingent tensions of existence are captured/understood from its dialectic. While the colonization of the inner planet is taking place, understanding these facets of the bodies/emotions as their "cobordant" instances (*sensu* Thom)[1] allows us to understand their limits and analytical possibilities.

The history of the theoretical, epistemic, and critical place of the alluded modalities could be considered as the very history of sociology, as attested in Alexander's four-volume work *Theoretical Logic in Sociology*, in *Modernity and Self-Identity* by Giddens, or in Taylor's *Sources of the Self*, just to mention three very different works.

Alexander states that:

> ... the epistemological perception of action abstractly considered to the specific consideration of action in society as studied in what are today called the social sciences, the concepts of subjectivity, objectivity, voluntarism, and determinism take on a character that is different from the meaning of these concepts in what is conceived to be strictly philosophical debate. In terms of their social, or sociological relevance, the implications of these epistemological commitments are altered in two ways. In the first place, instead of taking

the individual as the unit of analysis, the question of the nature of action must take into account the fact of the social interrelationship of a plurality of individual actors. Second, in terms of evaluating the nature of action itself, the problem of freedom and constraint becomes not merely a question of acknowledging the independent existence of internal motivation per se but rather an issue of the nature of that subjective factor itself.

(Alexander, 1982: 70)

As Giddens has pointed out, "the reflexivity of modernity extends into the core of the self. Put in another way, in the context of a post-traditional order, the self becomes a reflexive project" (Giddens, 1992: 32).

For his part, Taylor reveals:

I want to explore various facets of what I call the "modern identity". To give a good first approximation of what This Means would be to say That it Involves tracing various strands of our notion of what it is to be a human agent, a person, or a self.

(Taylor, 1989: 3)

It is not the intention of the present chapter to offer an introduction to the differences between self, I, and subjectivity (Wiley, 1994) since what is sought is to capture the tensions between disposition, position, and condition of concretion of the person. The interest in the person lies in the analytical and critical powers that it provides as a conceptual instrument to capture the experiences of human beings in the processes of colonization of the inner planet.

In connection with the above, it is necessary to emphasize that the pre-tension as emotional and motor predisposition to action that the dispositions designate are crossed intersectionally by the effects of positions and conditions in the structures as they have been developed in Chapter 4. Therefore, the individual/actor/agent/subject/author geometries and dialectics must be inscribed in the motion tapes, spirals, and networks that the structures generate as a condition of possibility of life-in-society that imply sociability, experience, and sensibility.

This chapter seeks to synthetically reconstruct the dispositionalities alluded to on the one hand as the geometry of the person, and as the dialectic of the person on the other, in such a way as to propose the emergence of dialectical personalism as a critical theory of the "new" coloniality that this book analyses.

5.2 Geometry and dialectic of the person

This section seeks to present the individual/form of an argumentative spiral, actor/agent/subject/author in such a way that they become present: (a) as a set of proximities/distances through which the situation of human beings in interaction can be understood, that is, as variable geometries, and (b) as a result of the processes and flows that correspond to the interactions between the aforementioned modulations in terms of their dialectical constitution. This geometry of the person must be inscribed in the dialectical tensions between the individual body, the social body, and the subjective body on the one hand, and the one between the body skin, the body image, and the body movement on the other.

The horizon of understanding should be drawn at the crossroads of what has been developed in the previous chapters and the introduction, since the presentation made here of dialectical personalism is focused on enabling a better understanding of the colonization of the inner planet.

5.2.1 Individual

The individual is the result of disputes and encounters for the production, distribution, and appropriation of basic nutrients to keep the nervous, immune, and endocrine systems in balanced articulation, ensuring the presence of human beings in interaction conditions. The individual is the organic unit that embodies the cultural/biological crossroads that sustains the species associated with the history of human beings. The individual is the result of the tensions between the phylogenetic and ontogenetic that make up a module of the organic as an unstable equilibrium; the individual is a complex system that is molded, among others, by the nervous, endocrine, and immune systems.

The central nervous system is made up of the brain and spinal cord, which serves as the main "processing centre" for the entire nervous system and controls all functions of the body. The brain involves a dynamic and plastic balance of chemical and electrical processes. The endocrine system, also called the internal secretion gland system, is the set of organs and tissues in the body which secrete a type of substances called hormones. Hormones are the body's chemical transmitters involved in many functions and circulate through the blood between organs and tissues. The metabolism, growth, and body development are processes where hormones intervene. For its part, the immune system protects the body from external agents such as bacteria, viruses, fungi, and toxins (chemicals produced by microbes). It is made up of different organs, cells, and proteins that work together and can be described in two main parts: (a) the innate and (b) the adaptive, which is acquired when the body is exposed to microbes or the chemicals released by microbes.

These processes are the result of the exchanges that human beings have with other living beings, the planet, and other human beings: the body/emotion is the result of social interaction. The social history made into a body is the basis of what we experience as isolated, but which is in fact a moment of the constitution of the relational. The brain is the most social organ of the body, which in turn is the most cultural entity in nature.

It is the organic unit that embodies the biological/cultural crossroads, which sustains the species, associated with the history of human beings. The individual, then, is an organic unit, the first reference of the individual, is what human beings feel as body/emotion. Here it is important to return to what was stated in the first chapters of this book about the body, skin, image, and movement, on the one hand, and the individual, social, and subjective body on the other.

The body as an organic unit is the first experiential reference to make said body intelligible. This intelligibility is connected with the fact that touch has to do with the sensory terminal units, this organic unit embodies a history, embodies the crossroads between the cultural and the biological, of what it means to be human, because the human being does not refer to the pure organicity, but also all the processes of this that we call the cultural/biological; it is in this same direction that in a theory of the social person, the five sides must be taken into account.

This organic unity is the result of human natural history, that is, the individual is inscribed in the development of the phylogenetic. We have a history of development, in organic terms, for example, the development of the hand, the development of the thumb, the development of the hand with the brain, the plasticity of the brain from historical processes, and the epigenetic consequences of the transformations in our body.

The history of the human being is usually told through the figure that begins in the monkey and reaches an upright human, taken as a metaphor for how the human being was taking an upright posture and developing the hand to grasp; it is precisely these exemplary qualities that speak of the genetic as archaeological, as phylogenetic; they also speak of the genetic as constitutive, they have an ontogenetic side, that is, how and what human beings are made of: our nature is on a par with the nature with which we coexist and interact.

Because all nature is going to be the result of interaction with the human being, but also with other living beings, and because that nature has been created by other living beings, we arrive at that nature with that presupposed interaction, or the other living beings that reach that nature, also presupposing our interaction with it. The history of the biological, of nature, can only be told by human beings, as we understand it, without denying, of course, even that there may be other languages with which to tell it. The biological/cultural character of that human history, which is a natural history, is a history of interaction with nature, it is a history of interaction with other living beings and the rest of the planet that remains as a trace of that relationship.

The individual is the first moment of the articulation of the metabolism that connects the social-natural and the natural-social, in terms of the manifestations of the energies of earth, water, air, and the senses to connect between human beings, living beings, and those energies including bodily energy. This socio-natural existence of being with others in the world as an individual carries in its most specific centre the question why to face others, it is in the flow of that question that the actor is born as another feature of the dialectic of the person.

5.2.2 Actor

Within the framework of the politics of the senses that was analysed in Chapter 3, the appropriate ways of seeing, smelling, tasting, hearing, and touching are outlined by the actor as another moment of the person in direct connection with the colonization of the inner planet.

The actor is configured in the interpretative possibilities of the action scripts that imply the accepted and acceptable politics of sensibilities in their orthodox, heterodox, and paradoxical modulations. The actor is the individual who plays a role arising from the construction/writing of a "script" of social relationships.

Interpreting implies the dramaturgical quality always oriented to a spectator/ another is where how to be, do, say, and feel is established by the social narrative contained in the script. Playing a role, a place in the scene involves where, when, and with what properties the character of the action is clothed, and what are the expectations of others in this regard. The social actor is built in the permanent tension of what must be performed, what can be presented, and what others expect to be represented.

The actor performs in various settings, fields, and spheres of dramatization where he represents specific forms of feeling practices associated with his position. The actor is the one who is going to perform a script on a stage according to a structure in certain conditions of existence. The grammar of the action varies depending on whether the actor's functionalist or dramaturgical approach is taken; the functionalist is closer to what we are calling here individual, and the dramaturgical closer to the modality of the subject because that personification will lead us to discuss autonomy more concerning dependency, and the first to think about the connection between function and goal.

Both the functionalist and the dramaturgical gaze make it possible to think of the distances between sociability and experientiality as keys to carrying out a definition of the situation (*sensu* Thomas) and the socially relevant existence of adjustments and misalignments of the action according to expectations. It is in the context of tensions and pretensions that the fact that the individual in a setting, and according to a certain social structure, is going to execute the performance that transforms him into a social actor is verified.

A question arises here: what does it mean to play a role? Executing is doing according to criteria before the action, it is developing an activity in seriality pre-determined by previous rules, and involves carrying out a practice according to previously established measures. Rules, criteria, and metrics of the expected behaviour of the actor are inscribed in the politics of sensibilities that allow unnoticed "incorporation" of them. From the point of view of performance, an actor performs, interprets, and represents an action, a text, a set of expectations about a framework of positions, and conditions of existence.

In this context, executing and performing are directly linked to practices attached to systematicity and automaticity. The actor is an individual for whom performing implies automaticity according to what is socially desired for that role and serially executing the action.

It is in this texture that a woman is expected to be a mother, a man to be a father, which highlights the discussion since the end of the 20th century and the current urgency about developing an intersectional and post-intersectional view of genders, ages, ethnicities, classes, etc., in terms of radically critical thinking, with the set of "it is expected that you" that makes the individual body as an actor. On the contrary, today society begins to see as unacceptable that the human being blindly does what society hopes that he does, with the obvious risk that this same rejection becomes a new rule to execute.

With what has been stated so far, it is clear that the actor is someone who does what the role tells him to do. Will and motivation are themes that are lost on the functionalist actor's side and partially recovered on the dramaturgical actor's side. This is so in a general sense because there is also another way of seeing this performance, which is the idea of representing, and this is getting closer and closer to the dramaturgical idea of the actor, who in dramaturgical theory will end up being a person, but basically what is represented is a role, a set of characteristics of a character, that is why person and character are not the same. The actor makes present traits of a character that embodies a typical character, that is why the script for the scenes characterizes what cannot be absent if a specific social person is to be present.

Let's put into action what it means to represent, for the actor's logic, three things: (1) to put on stage, (2) to present qualities again that is why it is to re-present, and (3) to witness some qualities in the middle of that stage, for example, the soldier, not only returns to put the presence of the monopoly of physical violence of the State, but also returns to stage someone who is performing, so to play means to represent, he is in the place of honour, bravery, the old distinction that Plato made of social classes that depending on where the soul fell, on the chest, on the arms or the head, one was going to be a philosopher, farmer or warrior, then it is the characteristics of those souls that make the subject have to be that whole life. This is what clarifies what is expected of each human being and what their claims should be.

These are these three ways of entering the logic of representation, of putting the present again, of recovering the characteristics and this idea of "I am here on behalf of" is a practice that human beings play certain roles to coordinate action, adjust expectations, and reproduce society. Digital education is a radical example: an actor plays a role of a teacher/speaker, then, for each of the participants, one behaviour is expected in a way and represents it in a way, with the tone of voice, the enunciation, the "language in use"; it is not necessary to see the face, then there are a series of structures where representing implies something of what it means to perform, which obviously cannot be represented without executing, but this representation is based on these characteristics that the teacher takes, owns, and assumes that class participants will do the same.

The foregoing indicates in a sense that leads to another facet of playing a role, which is that of incarnating, on the horizon of oscillation that seeks to mark between function and performance: the actor makes a body of certain rules, therefore it is possible to see that each of the social actors have a different corporal hexis (*sensu* Bourdieu); the intellectual, the priest, and the executive, they stand in a way, they speak in a way, then these are also dictums, imperatives, and impositions of the social structure, of how to embody a role.

Human beings are not a kind of result that is quite dissimilar to foresee and that is why "we must equip" each one with the possibility of representing roles in different ways, precisely in the distances and proximity between executing, representing, and embodying.

One more step to understand what this dialectic of executing, representing, and embodying means is possible by investigating the script that is being acted out. The notion of the script is clear for structural functionalism, neo-functionalism, and also for dramaturgical theories, in which these five basic characteristics of the action are present: (1) it has a goal, for example, the soccer player has one, the artist has another, the university professor another, and the teacher another; (2) it has rules, like "sens du jeu" (*sensu* Bourdieu), knowing how to play football, explaining theorems, etc., – there are rules by which people know how to perform, how to follow the rules; (3) the other actors expect that as a specific actor will play an action, a sailor knows how to navigate; (4) there is an expected result of the action, for example, the tennis player has one result and the soccer team another, one is an individual game and another is a collective game; and (5) there is also a narrative, in the sense that "I believe that my role is this."

It is in the tension of becoming an actor, an individual, that the flow of sensibilities leads to the context of the ability of human beings to be able to do what has been done again, to amend mistakes, to inspect what has been done: this is how the agent side of the person appears as a specific moment of their existence.

5.2.3 Agent

The agent implies a turning point from where the contradictions and paradoxes that emerge from the abilities of human beings appear in a game of proximity and distance with the conditions of the colonization of the inner planet.

The agent is characterized by his ability to re-do actions in terms of their consequences and intentions, managing resources and goals, a skill that is anchored in his ability to understand the action as a result and as an input for another action.

The agent is the empowered (right-handed) individual and actor with skills and abilities to elaborate on the social "texture of" and the "text of." The texture operates as an intentional instantiation of some features of the materiality of the action and not others, being linked to the tension with what is individual in it. The text is the ability to re-assemble the script that the actor has, whose disposition connects him with what we will later understand as an author.

Skill is a way of exercising the ability to be in co-presence with others according to differential margins of potentialities as a "frame" of doing things with power over and power of. Power of domination, of elaboration, from top to bottom; or the power to build with, to do between, and to co-create the world.

The agent has skills, powers, and abilities. The skills are the moments in which it carries out, with a certain effectiveness and efficiency, particular tasks, be they of social interaction with other human beings, of interaction with processes, with institutions, or with other living beings with which they share the world. The agent "managed" the horizon of understanding that have the persons as inhabitants of the social world, a "subjective" world, and a "natural" world.

These skills are capacities that the agent has, which is a possibility, which arises from the tension between his traits as an individual and as an actor, and the "plus" that he embodies as an agent. That is why in sociology, class theory, stratification theory, and function theory, are associated with the logic of designing, producing, reproducing, making known, or teaching skills.

A first clarification is that these skills are differentially and unequally distributed according to geopolitical and geocultural sensibilities. In the emergence of capitalism, the worker "corresponded" to position his body as a machine, there the logic of power is played; therefore, the unequal distribution of skills and powers has to do with the capacity of those individuals who cannot be either agents, or subjects, and even fewer, authors; they are in the structure as mere actors to reproduce or not reproduce the social. Depending on the time/space that human beings inhabit, we can all be agents, but not in the same way; we are all agents, but in a different position, disposition, and class condition. The horizons of existence are those that allow the acquisition and exercise of powers and skills. The power has to do with this constitution of the subject to be able to carry out the skill in which the subjects begin to have possibilities; therefore, between the skill and the power, there is the ability of the subject now understood as this possibility of carrying out a consistency between your power and your skills.

From the moment of primary and secondary socialization, the construction of institutions and the re-formulation of the cultural systems of needs and the logic of construction of the future, beliefs, skill, and power, are essential parts of the creation of an agent and the power dispute is the possibility of realizing stabilities that are unevenly distributed. So, there are a series of inequalities that begin to emerge, the inequalities that come from functional illiteracy that have one at every moment in the history of capitalism: first you had to have strength, then you had to know how to add, read and write, and now, to be in the digital age, you must be familiar with devices and applications.first,

If the historical chain of the reconfiguration of powers and skills is noticed, what is found are these logics of making and re-making the world, and that is what we award to agents, therefore, in societies normalized in immediate enjoyment through consumption, such as ours. In these societies, inequality and the distribution of skill and power are eminently associated with the origin of gender, class, age, and ethnicity; they are great drivers of inequalities, which are perceived by individuals/actors, they live them, they transmit them as insurmountable given their perception of the impossibility of completing, at least partially, a "metabolism of equity" with all living beings.

From another perspective, it is interesting to note that for the agents there is a texture of the action that is connected with what was analysed in the first chapter as individual, social, and subjective bodies in terms of social history made body. The agent is embodied, made a body, "incorporated," there is a construction of corporal-affective logics, for this reason "embodied" becomes flesh in the sense in which the agent emerges into a different body.

If you map the world of hunger and need, and the world of luxury, of what is left over, of excess, agents with different bodies are found because the social is incarnated. Agents have different hands, different mouths, skin types, weight lifting abilities, or walking balance. In this framework, the embodiment can be understood in three ways, but it has a succession of processes; the making of body, of social history, is to constitute the corporal hexis, which has to do with how it has been incarnated, with what type of body it has taken shape. That is why the place of the law, of the rule, of the norm, and roles, have become a body, they have been incorporated when incarnated, they "move away" from the actor, giving way to the agent.

When a body/emotion is seen, the social is observed; it is not that the body is a representation of the social, it is social, that is why the agent is the social made body. That is why the agent is of interest as a carrier and builder, because it "carries" a skill to separate the logic of production and reproduction that nests in what we call the body, which as such is an epigenetic framework, which connect with the social history made body, in-embodied, that is to say: with the history of the tensions between individuals/actors/agents.

In tension with the above, the agent, too, must be understood as the "origin" of the dispute over non-agency, or denial of agency, of losing the faculty, capable or incapable, the "not for me," "not being able," being afraid, not knowing, which reaches the logic of not wanting, the impossibility of wanting and the relationship between desire and need; it is precisely why agents are the vector space through which the dispute over being autonomous takes place.

It is individuals/actors who have embodied the permanent dispute over autonomy in this process of permanent incorporation of social forms; this dispute over autonomy, or its opposite, will later be present in the subject trying to build itself through a permanent process of the search for autonomy.

Thus, the agent is a side that "co-borders" (*sensu* Thom) with the subjective, and "confronts," in tension and vectorially, with the individual and the actor. Placed in opposing pairs and in solidarity with each other, they are possible to be understood as the tensions between individual/actor and agent/subject.

For this reason, what sociology of bodies/emotions seeks, primarily, and not only, is to establish what are the forms that this process of affective cognitive incorporation takes, in such a way that it becomes unnoticed and alters the ability and dexterity of the subject to perform the action.

5.2.4 Subject

The subject is the reflective capacity of appropriation of the distances and proximities between individual, actor, and agent that provides the possibilities of identity, autonomy, and encounter. The subject is the self-experience and self-competence of "holding together" the potential styles of actor and agent from a material-reflective "point of experience" in tension with the traits of the individual/actor/agent.

This notion of the subject recovers the tension that appears in Marx between consciousness, consciousness in-itself, and consciousness for-itself, this modulation of reflexivity has characterized the processes of the dispute over and for autonomy for several centuries. The fundamental difference is that it is the notion of experience of the sensible that will guide reflexivity in opposition to the enlightenment and rationalist view of the power of consciousness.

The idea of self-experience implies that the subject feels "in a situation of experience," and the individual/actor/agent are moments that are stressed with said experience; the subject has to do with the competences of keeping all the ways of being a person together in a centre of gravity; therefore, the subject is the one who binds, the one who holds together, the experience of the personal, because it is a way of living the connection between act, activity, and action reflectively. It is this "feeling-thinking" (see Introduction) that starts from the biological-cultural nucleus and reaches self-experience. I feel therefore I am; I love therefore I exist, replaces the logic of the pre-eminence of consciousness as thought.

What is it that holds together, what is it that draws this centre of gravity? It is the possibility of designating oneself as responsible for the modifications of the social, subjective, and "natural" world. The subject is the result of the diversity of ways of being individual, actor, and agent, but, from an experience point of view, where sociability and sensibility can be appropriated from the agent's skill, according to certain rules of the actor and in connection with the material of the individual, so that it is experienced as a constant flow, as a structural contingency, and as an iterative moment.

Cognitively, the subject displays his ability to capture units of experience, because the subject provides the paste, the cement that he extracts from how the experience is lived, and the individual/actor/agent has to do with how that experience is forged, whit the relationships between nutrients and possibilities of coupling with the world, how are the possibilities, powers, and deficiencies of connexions with living beings and that interrelation with what is called natural, which is nothing more than the "history" of becoming human and his relationships with the planet.

After all, there is no story if it is not told by these human beings, there is no such narrative without nature, nor nature without the practices that make it possible and/or deny it. That is why it is a material/reflective point of experience, that not only comes from the individual, but it is questioning what and who I am. It is in this vein that it is elaborated, in a subjective moment, that implies, one space of the subject, a tension between the self and I; since the subject is precisely about constructing the I as an experience and as a centre of gravity.

The agent has to do with "running/management," the actor has to do with performance/management, and of course, it is epigenetically related to the individual and the subject connects with the *experiences* of these tensions. The history of subjectivities has to do with the potentiality of the individual's history but from the point of experience of what has been reflective and what has constituted a personal experience.

Why is it epigenetic? Because attached to what Bourdieu developed about habitus, human beings constitute their bodies/emotions in and through history as present, and the future as an acted out present. Thus, a paradox is instituted: the weight of the structure on the subject that, in reality, follows and produces the habitus, then, the relationship between reproduction and production is always a history of the dialectic of subjectivity as an appropriation of experience.

For this reason, one can speak of historically determined subjectivities, because the epigenesis of these subjectivities has to do with previous structures, with the structures that others have experienced. The subject is the first-person carrier of the story reproduced and to be reproduced, it is an experience of subjectivity.

Returning to the dialectic: the actor plays roles, the agent is skilled in re-doing the action, the author creates, the individual carries the biological-cultural nucleus, and the subject unfolds an experience. The connection between the centre of gravity and experience refers to the fact that it is the subject that ties together the social person as an identity, it is subjective experiences that "tie together" the modalities of the person in a dialectical way. As anticipated in the Introduction, dialectic means tension, but also propaedeutic, because one way of posing the subject is that it is a propaedeutic of the other, to the others. That reflection is what allows one "to realize that I am I," there would be no possibility of there being another if the history of subjectivities did not exist, that is why subjectivity is the appropriation of these stories that human beings have as experience and as a centre of gravity. For this reason, the set of intersectional dispositions is what builds your intersubjectivity, which is the "I am a subject," that would be a third question that we have advanced. The self is a centre of gravity insofar as it coordinates the positions, conditions, and dispositions for action; because it says them in the first person: "I love," "I wish," "I do," they are dispositions, positions, and conditions that have to do with the history of that centre of gravity as a subject.

The vision of the subject is a vision that brings into play, that articulates, that enables, and also makes impossible, the results of the subject's positions in the social structure.

In this relationship between position and condition, there is no subject without family, without peers, without institutions, and that is lived in a classed, ethnic way, by gender: there is no subject without groups of references. The modulation of the subject is the spiralling process that is played between reflexivity, autonomy, and identity that opens the door for authorship.

5.2.5 Author

The dialectic of the person has its last moment in the author, where the break with the colonization of the inner planet becomes a central point. The author takes up the agent's abilities and the subject's autonomy in a disruptive attitude in and through creativity and expressiveness.

The author is constituted in the creative and expressive possibilities of elaborating paintings of the world as criticism, re-design, and extension of the politics of sensibilities. The author is the possibility of changing the position and conditions manifested in a special and particular way(s) of doing, creating, and expressing.

The author, unlike the previous modalities of the person, is not a quality of the individual or subject, or a capacity of the agent, but is a position and a disposition concerning a condition. A social person is a form and a dialectic, this dialectic of the person implies that in a moment there is a possibility that the one who embodies the logic of action, becomes the one who writes, combines, and structures the same action. The most important feature of the author is that he can change because he can re-make, re-write the paintings of the world of sensibilities, sharing, in part, this feature with the agent.

The definition of author not only focuses on his doing, the doing has conditions and positions that make it an originator of grammars of the classed action; but also, the author is the one who works/operates in the logic of the disposition, because he has a connection with what that individual/actor/subject/agent does as a dialectical totality. The author is the one who provides a narrative experience sharing this trait with the powers of the subject.

Another skill is also rooted in the agent and has to do with making the world, thinking of the world with this tension between the subjective world (subject-agent), the social world (characterized by the presence of others in the actor-individual), and as the natural world, the others that inhabit the planet as the horizon of creation.

It is in this framework that the "coexistence" of doing, expressing, and creating, in certain conditions and positions, must be noted; in terms of agreement concrete possibilities, which are agreed on specific possessions. That is why what the author will end up arguing/disputing are these dispositions, these capacities that become the conditions of the possibility of a disposition.

The key to this look is the notion of possibility, and to clarify it, it must be inscribed in the subject/agent/individual/actor tension. To see the attribute/quality of an author, what he creates, what he does, and what he expresses, it is necessary to return to the relationship between skill, power, and the agent's ability. How the agent is to put/be in the world in such a way that, being right-handed, he can redo the action. There is the possibility of creating, of moving to the creative act "in-action," because it can be re-made. It is not possible to perform the "same acts," but to redo the action, for that we have this tension between power, ability, and dexterity. There is a "non-determination" of the agent and the subject, and this is not due to the obvious presence of the material conditions of existence, but because there is a possibility of applying the skill again.

Because the author is nothing more than a tensional form of the subject, where the autonomy of the subject is expressed in such a way that it works/operates from its point of experience, taking up again the notion of the subject as a point of experience, as an analogy of the point of view.

The point of experience is the point of autonomous experimentation that human beings have as subjects, but which is expressed, as an author, in the ability to make their script. In connection with this, it is important to return to another moment of the dialectic: the individual not only responds to the reproduction of his biological side, being a tributary of both neuronal plasticity, the central nervous system, and the immune system, but mainly as a result of the biological turned cultural and the cultural turned biological. Achieving increasingly autonomous spaces of points of experience becomes the key to a dialectic of the person as resistance to the colonization of the inner planet.

To achieve the above, it is necessary to elaborate a look at the moment where the person is the author of the action; authorship and autonomy are concomitant traits. Criticism of the unequal distribution of nutrients may or may not become a way to appropriate this inequality, in terms of subjective autonomy, the skill of the agent, and the possibility of the author to write other feeling practices. For human beings, writing our biography has to do with how we lock and unlock the texts written by the agent-author, reproduced by the actor and inherited by the individual.

In the individual-actor tension, we are playing the step to the possibility of autonomy or the possibility of skill, and for that reason, for us, the material conditions of existence enable a willingness to redo the action and write new scripts as agent-authors.

Within the framework of this dialectic, it is possible to rethink social determination: this conditioning leaves the agent stiff, and a frozen agent is precisely a "non-subject" who has become a mere actor who reproduces a libretto written by the society, with its own individual traits linked to the phylogenetic and its class conditions and positions that it cannot transform; then it is necessary to break the vision of society only as a photo, a photo of the never-ending, a photo of reproduction.

Seeing society from the agent and the subject who can transform that condition and that position in a disposition, he becomes an author who rewrites that book and rewrites his biography. Returning, an author is the one who makes, creates, and expresses. This logic of doing and undoing has to do with two very important elements: doing is the capacity that we have, the possibility that we have, to link the agent, the subject, the individual, and the actor. This capacity, this skill, this ability, and this autonomy of the experience, means transforming the disposition that we have in front of the conditions and possibilities and the position. Doing transforms, the condition and position; the condition is the elements' material conditions of existence, the position is the place in which the human being is inscribed in a grammar of the action and geometry of the bodies, of which the individual-actor relationship is a tributary.

The author is a performer and scribe who re-arms the subjective, social, and "natural" world inscribed in the tension of dialectic and geometry of the person, a tributary of a geoculture and geopolitics of expressiveness.

The author is the one who "counts" the planet and the human beings themselves as part of that planet; which is nothing more than the story that we tell about how we are part of that planet. In a preliminary sense, we are the planet because we are the ones who tell the story of that planet, that is why the predation is so radical that by predating the planet, we predate ourselves: we colonize the inner planet.

The main predation is of the individual energy and of the corporal energy that is ready to do, precisely, what we human beings do because we have that energy, the general planetary predation is predation of the body and the corporal energy-individual.

It is the author who not only writes the narratives, the imaginaries, and the ideologies, but he is the one who is going to narrate, precisely the sensibilities; it is in this praxis that action becomes narration and narration becomes action. That is why speech acts, performative consequences of speech are performative consequences of the action. What we do is reproduced in the other actions, so when we discussed the notion of the subject, we saw that there is a link in subjectivity, between recursivity and reflexivity. That is why doing is thinking about doing, and doing it as we have thought about it, that is why representing the world is already intervening in it. As expressed in Chapter 3, social practice begins with sensations, and it is from the occupation and management of those sensations that the colonization of the inner planet deals.

For this reason, doing is intervening in the world: when human beings represent the world, they are doing it; this is not any form of subscription to the idealistic variants, nor romanticism, nor miserabilism, nor populism at the theoretical level. But a rescue that practical logic is theoretical logic, what transforms us into authors is that when we do, we think, and when we think, we do; therefore, all action implies a praxis. In the context of the dialectic of the person, the human being is taken up as the author, which implies a point of experience that imputes sensibility to the distance between condition, position, and disposition.

5.3 Existence and sensibilities

It is predation on a planetary scale of all types of energy (mainly body) that builds the network of mimetic consumption, resignation, and diminished humanism, in its spiralling process with the consecration of societies normalized in immediate enjoyment through consumption that is cemented in the Logic of Waste, the Politics of Perversion, and the practices of Banalization of the Good; opening up like another Moebius band in the sensibilities of platform and looking-touch at the playing of Society 4.0 are the scenarios of the dialectic of the person that involves the individual/actor/agent/subject/author.

The dialectic of the person allows us to understand a little more and criticize the complexes of the colonizer that are enunciated in the Introduction of the book. Pontius Pilate is the indifferent colonizer who "naturalizes" the occupation as a concrete historical way of instantiating a destiny embodied in points of experiences such as "the exploitation of rare earths is an inevitable fact of development." Columbus is the colonizer who elaborates his point of experience on the occupation from the benefits that it brings to the colonized/conquered – "the exploitation of rare earth has great benefits for the human being." Nero is the usurper who appropriates what belongs to others, by socializing the experience of superiority of his actions as the only option – "the exploitation of rare earths is the best thing that can happen to human beings." The dialectic of the person makes it possible to criticize these complexes of coloniality and reflectively stress the traces of an inquiry that denies the supposed truth value of a political economy of morality based on a politics of the sensibilities of the colonization of inner planet.

This geometry of the person is stressed and shaped as a mode of existence and criticizes the political economy of morality; it involves an alternative sensibility that challenges the acceptable and accepted politics of sensibilities for its claim as a single and closed totality. To exist, to live, and to sensitize are the three moments that emerge as a consequence of the dialectic between the appropriation of the phylogenetic and ontogenetic that the individual carries, the criticism of the sociabilities and experiences that the actor must interpret, the empowerment of the agent's skill in re-doing the action, the autonomous reflexivity of the subject who sees himself as the axis of his identity, and of the collective and the author as the creator of new possibilities of making a life.

Both the geometry of the person and their dialectics leave us at the doors of a final discussion: what are the priorities/values and how can they be understood in societies where the colonization of the inner planet is increasing day by day? In a first approximation, it would seem that individual/actor/agent/subject/author in his dialectic should make us think of the common as an entrance/exit door to the paradox of colonization carried out on a planetary basis, but on the scale of that geometry.

In the context of the geometry and dialectics of the person, what is exposed as structure in Chapter 4 and the processes of social reproduction and the situation of conflict networks involved in this, the next chapter explores collective actions as forms of social practices.

Note

1 CFR Thom (1952, 1976, 1977, 1982, 1989, 1999, 2008).

References

Alexander, J.C. (1982) *Theoretical Logic in Sociology, Vol. 1: Positivism, Presuppositions, and Current Controversies*. California: University of California Press.

Giddens, A. (1992) *Modernity and Self-Identity*. Cambridge: Polity Press.

Mbembe, A. (2011) *Necropolítica seguido de Sobre el Gobiernos privado indirecto*. Barcelona: Melusina SL.

Taylor, Ch. (1989) *Sources of the Self: The Making of the Modern Identity*. Boston: Harvard University Press.

Thom, R. (1952) "Espaces fibrés en sphères et carrés de Steenrod." *Annales scientifiques de l'École Normale Supérieure*, Sér, 3, (69), pp. 109–182.

Thom, R. (1976) "Crise et catastrophe". *Communications*, 25, La notion de crise. pp. 34–38.

Thom, R. (1977) "Structural stability, catastrophe theory, and applied mathematics: The John von Neumann Lecture", *SIAM Review*, 19, 2. (Apr.), pp. 189–201.

Thom, R. (1982) "Teoría de Catástrofes y Ciencias Sociales una entrevista con René Thom", *Entrevista de José Luis Rodríguez Illera, El Basilisco, 13, noviembre 1981-junio*, www.fgbueno.es

Thom, R. (1989) *Esquisse d'une sémiophysique : Physique aristotélicienne et théorie des catastrophes*. París: Interédition.

Thom, R. (1999) "AVANT-PROPOS. Epistémologie et Linguistique." In: Wildgen W. *De la grammaire au discours, une approche morphodynamique*. Bern: Peter Lang.

Thom, R. (2008) "Prevedere non è spiegare. A cura di Giuseppe De Cecco, Giuseppe Del Re e Arcangelo Rossi". In Dipartimento di Matemática (Ed.) *Quaderno 3/2008*, Lecce: Dipartimento di Matematica "Ennio De Giorgi" dell'Università del Salento, 2008.

Wiley, N. (1994) "History of the self: From primates to present", *Sociological Perspectives*, 37, 4 (Winter, 1994), pp. 527–545.

6 Collective actions

6.1 Introduction

Chapter 4 argued and underlined how it is necessary to reweave and re-string the connections between structure, process, experience, and emancipation, in the context of a Moebesian tension: the internationalization of the emotionalization regime, the structures of compensatory consumption policies, and the political economy of the morality of public infrastructure.

The previous chapter exposed the basic features of the geometry and dialectic of the person: individual/actor/agent/subject/author in a spiral where dependence and autonomy mark the experiences of the coloniality of the inner planet.

In this context, this chapter seeks to point out some analytical clues to connect conflicting structure, emotions, and collective energies within the framework of the process of colonization of the inner planet.

As I have maintained elsewhere, conflict is not only a disruptive social fact; it is a piece of the whole building of society. (Scribano 2004, 2012b, 2019) The riots, silent marches, and all new aesthetics of protest show and tell us about identity, differences, and the fragmentation of society. We have the challenge of looking carefully into the possible interlacing between phenomenological aspects and structural features of social protest and their multiple connections with sensibilities. To interpret the meaning of collective action, we need to pay attention to three components of social protest and movements:

1) We need to explore collective actions as expressions of absences that point out the moments when the social system could not be sutured, when society has no cement for linking social practices. We need to find out the mechanisms to express the absences in the social structure and the processes of absenting absences. That is to say, the path through which collective action shows how social reality is constructed over the social fault.
2) We need to understand collective actions as symptoms of social conflict, as a sign of social relationships. Social protests are a manifestation related to social conflict and the functions that it has in the social organization.
3) We need to interpret collective actions as messages that mark the borders of systemic compatibility and tell us about the "limits" of social power.

If we put together the critical analysis of the absences, symptoms, and messages that talk about collective actions, it is possible to see how our own theories are inside the field of the Academic Colonial Fantasy, taken as an ideological mechanism and a politics of academic sensibilities. Finally, telling a story of social protest involves becoming a part of the conflict network or, at least, contributing to building one side of hegemonic discourse. Writing this story is a practice of identifying and cowriting a social scientific account about social protest messages.

Reflexive skills may be used in this intersubjective position. This scientific practice implies (paraphrasing Melucci) listening to and interpreting the practices of the nomads of the present.

Harriet Martineau, who was the founding mother of sociology, in her important book on the methodology of social research, *How to Observe Morals and Manners*, wrote:

> The need of mutual aid, the habit of co-operation caused by interest in social objects, has a good effect upon men's feelings and manners towards each other; and out of this grows the mutual regard which naturally strengthens into the fraternal spirit.
>
> (Martineau, 1838: 219)

In the context of the colonization of the inner planet, it is imperative to provide a look, at least preliminary, at how collective actions are structured in terms of effects, responses, and contexts.

The first approach to collective practices is to point out the existence of emotional ecologies that accompany them on the one hand, and structure them on the other.

6.2 Emotional ecology

It is in this framework that the recognition and critical analysis of the emotional ecologies that we have at hand acquire importance,[1] which can in some way help to relocate the pieces of the game, which will be beyond whether we accept their presence or not.

An emotional ecology can be characterized by three factors: first, in each politics of sensibilities, a set of emotions are constituted connected by aspects of family, the kinship of practices, proximity, and emotional amplitudes. Second, this set of emotions constitutes a reference system for each of these emotions in a particular geopolitical and geocultural context that give them a specific valence. Third, they are groups of feeling practices whose particular experience regarding an element of life can only be understood in its collective context.

In the first sense that we are pointing out, an emotional ecology is being constituted by those emotions that are in a similar chromatic field. With sadness, melancholy, and anguish, for example, we are forming a surface of emotional inscription that allows us to understand the content of each one by the relationship of proximity and distance that each one acquires in the field/space that is formed on this surface. Joy, happiness, and joyfulness offer another example of how, in a given society, they can be understood through the proximity and distance in which practices acquire their experientiality and sociability. These aspects of the family allow emotion to occupy a place in the field, given a certain value of attraction and rejection with another that inhabits that same ecology: immediate enjoyment through consumption means that happiness and joy are experienced differently, but are in mutual reference. They are kinship to practices that, to be captured, must be put into play in the identification and assessment of each one and the whole. Enjoyment can only be explained by accepting the differences and similarities with joy, happiness, and joyfulness about consumption.

On the other hand, emotional ecology refers to the weight of where and for whom this set of practices taken as a whole is lived. Thus, there are the political and cultural valences of what can and should be felt in association with each of these references. The scenario constituted by the politics of sensibilities is conditioned by the spatial distribution of power, its territorial organization, and the borders and "bridges" that unite/separate the practices of feeling. It is in this sense that an emotional ecology must be understood within geopolitics that provides the parameters for experiencing emotions in particular. In a similar direction, an emotional ecology is structured based on the cultural identities and particular ways of life of those who experience those ecologies. The unequal distribution of nutrients, the differential access to sources of bodily energy, and the inequality of possibilities of "eating healthily" are the manifestation of how the geopolitics of food conditions the experience of the anguish of scarcity, social suffering in the face of not eating, and the "heaviness" of full bellies. In this case, it is also palpable how an ecology of fear is detectable in war zones, in migrant and refugee camps, and in the daily lives of women in the face of femicide; regions, countries, and continents that are geopolitical structures of an emotional ecology.

Third, an emotional ecology implies the collective imputation of the experience of a set of emotions concerning processes, people, and objects, that is, emotion is performed from the collective socially learned experiences, its valences, and chromaticity in connection with a specific element. Sadness, anguish, and pain in the face of death are constructed differently, sieved and socially organized. What to feel, how to feel it, and in what way to express it nests in pre-existing societal experiences that are apprehended and learned as a member of a collective. In the face of deaths, births, love unions, and birthdays, the connections between happiness, joyfulness, and joy are different. A life lived, everyday life, is marked by politics of sensibilities where words and things acquire volumes, densities, and values. It is where things and words are inscribed in one or another emotional ecology; from the insult to praise, from the photo to the TikTok video, and from the political slogan to the religious interpellation. Planetary emotionalization is the "glocal" result of a political economy of morality that harbours politics of sensibilities in which the diverse political ecologies nest.

In this picture of the situation, it is interesting to review that one of the sources that connect the structures, as explained in Chapter 4, and the appearance of emotional ecologies are precisely the collective actions and social protests, with which the next section deals.

6.2.1 *Social protest and conflictive networks*

In this section, it is shown that there is a direct connection between proximities and distances, the pluralities of conflictive networks that emerge in a social protest, the expressive resources used in them, and the social structuration processes in which this is inscribed. There are also well-known relationships between emotions, collective actions, forms of protest, and expressive resources[2].

Generally speaking, any collective action that becomes a social protest could be an indicator of one or more conflicts. Such conflicts can be defined in their primary and ontological quality as the result of the diversity of values that two or more agents can have about a good they deem to be important. This importance could come from the quality of the goods about the agent's material, from its symbolic weight, or from other mechanisms of social reality that alter the production, accumulation, and distribution of the goods indicating then a collective problem. In this sense, a social protest is preceded and presided over by conflicting situations. These conflicts can be referred to as conflictive networks.

The conflictive networks that precede and operate as a background for the protests change over time, reconverting and redefining the positions and sense of actions of the agents. In this sense, a protest can only be understood if it is analysed as the interconnection of diverse "moments" of mobilization that are generated and revolve around conflictive networks, but are not depleted by the mere manifestation of collective action; rather (networks) are strongly related to their inactive periods, an issue that will be explored in the next section.

An analysis of a protest should always keep in mind the conflictive networks, such as those that shape the "conditions of collective action." It is important to understand that such a "network of conflicts" is constituted by the relations between the actors and previous conflicts with which they are interconnected. Besides, it can be observed that a network of conflicts gives possibilities of social visibility to another network of conflicts, and that is potentially submerged in everyday social relations.

The connection between conflictive networks and social structuration processes can be analysed in terms of absences, symptoms, and messages. In this sense, it is based on three axes that serve as the guiding thread of the analysis: to understand the social protests from the absences, as symptoms of social structuration, and as messages of the redefinition of the limits of the systemic compatibility of a society. In a very simple theoretical framework as a means of "entry" to these conflicts, we need to understand the symptoms, absences, and messages, which come from these conflictive networks.

In the cases where they are symptoms of social structuration, they manifest. If there are absences, they reveal. In the case of messages, they communicate. In the *first place*, collective actions are an epiphenomenon of what happens in society, remembering that they always refer to conflict networks, and these networks are always related to the structural and to the structuring processes, whether in their material and/or "symbolic" moments.

The symptomatic works by metaphorical transposition; from a sign, it can be understood as a set of relations to which that sign has no direct reference but it implies it. In the *second place*, the protests evidence absences, gaps, and moments of social relations where social logic cannot unite the "natural" ties between agents. They also refer to failures in the social structure that generate cracks, places where the social structure has been broken and from where there are no longer bridges that link the broken parts. The protests show the places where society has no cement, where the pieces cannot be joined, where it is not sutured. In the *third place*, protests are messages because they essentially speak of the limits of systemic compatibility. In this sense, they manifest the state of the mechanisms of conflict resolution, they point in the direction from where society can no longer limit itself and where the borders have been trespassed, demanding then a systemic redefinition effort. Protests reveal the precarious relationships between the limits of compatibility and systematic incompatibility, sending signals from places where, if the system wants to push more, it cannot do so, except at the cost of its redefinition or dissolution.

Absences, symptoms, and messages can be understood/captured through expressive resources "put-in-play" and from the conflicting networks expressed in the protests.

6.2.2 Expressive and "Aesthetic-in-the-Streets" resources

From this context, it can be inferred that the modes of expression rooted in and based on the sensibilities of the subjects are tied to historical practices that seek a heterodox place for domination and should be read as a task of rehumanizing relations between humans and as politics.

It is obvious that the expressive resources are anchored in criteria and aesthetic valuations of materials, production, and artistic sense, and that these are in turn dependent on a particular historical context. However, for the expressive resources of collective actions and social protests, there is a making visible and perceptible what was previously invisible and inaudible.

Aesthetics becomes a politics of the senses, heterodox and an opener of worlds that become palpable in (and through) that same practice. These expressive resources are linked politically, and it is usually the task of the institutional policy to silence and sterilize the "here and now" of claims and demands, transforming them into elements of their policies. "Aesthetic-in-the-streets" then, is a way of jumping the barrier of the duplication of the natural as one of the axes of domination and "transcending" against the given.

These discussions took shape between inquiries, misinformation, and re-appropriations that, in a new scenario, could be questioned. Thus, in a monochromatic society where the capitalist system elaborates a set of devices for the regulation of sensations, it is very important to recover those readings where the aesthetic is linked to the possibility of configuring, showing and activating new sensibilities. The connections between expressive resources, aesthetic-on-the-streets, and collective action will allow us to dive into the aforementioned readings.

Expressive resources offer a double possibility for reading: they are constructed and used as products of the senses (results), and are at the same time senses in production (inputs). From the perspective of inputs, resources are selected and used to reframe their original position in a plot of new significance, while from the perspective of results, resources are sifted through a process of meaningful production, becoming thus a "novel" utilization.

This quality of products in senses and sensibilities-in-production is what gives mobility to reading, understanding those that could have re-meaning, and those that still persist and are being built. So then, they are results and inputs that allow an entryway to observe legacies, reconstructions, and new creations within the practices of the action, and the senses that the subjects give to the protest, visualizing problematic areas within the conflicting networks on which the action is built.

The expressive resources in this context can be better understood if the connection and relationship with the sensibilities are considered, and that these, in turn, come from the weave of sociability and experience. What societies and individuals "have at hand" at the "time of protesting" are the sensibilities not only as a product, but also as an "input" of the processes of social structuration.

The different types of families, the multiple ways of teaching and apprehending the various modes of doing justice, and the rules that need to be accepted in society, are some of the complex long-standing relationships that involve possible sociabilities. The different positions that each agent takes from the sociabilities, the amalgam of existential paths that each subject produces/reproduces from the institutional frameworks, and the contingent ways of being in such frameworks, outline the possible experiences in a particular society for a particular subject. The patterns and feeling practices, the politics of senses (what can/cannot be smelled, liked, touched, seen, or heard), and the practices of desire associated with the limits and potentials of sociabilities and experiences, constitute the sensibilities (accepted/acceptable) in a society.

As we have already said, social agents know the world through their bodies (senses). Impressions of objects, phenomena, processes, and other agents structure the perceptions that the subjects accumulate and reproduce. From this perspective, a perception constitutes a natural way of organizing the set of impressions that are given to an agent. Such configurations consist of dialectic in the tension between impression, perception, and the results of these, which gives the "sense" of surplus to the sensations. That is to say, that places them here and beyond the aforementioned dialectic. Sensations as a result and as an antecedent of perceptions, give rise to emotions as derived from the processes of adjudication and correspondence between perceptions and sensations. Emotions understood as consequences of sensations can be seen as the puzzle that comes from the action and effect of feeling or feeling-one-self. Thus, identifying, classifying, and turning the critical play between perception-sensations and emotions is vital for understanding the mechanisms that regulate sensations and the mechanisms of social sustainability that an asset has as one of its contemporary features for social domination.

The dialectical tensions between sociability, experience, and sensibilities that are "deposited" in expressive resources, are indeterminate and contingent but hermeneutically relevant results, that allow us to understand, at least initially, the principal features of a cycle of protest.

A new Moebius band is opened from the connection between expressive resources and sensibilities: there are collective actions that deny the value of a closed totality that, as we saw in Chapter 3, the politics of the senses of the colonization of the inner planet claim to have. It is in this context that our interest in interstitial practices increases and is presented as one of the axes of this chapter.

6.3 Interstitial practices, collective interdictions, and experiences of affirmation

From various investigations, we have been insisting on the need to consider a set of social practices that allow us to look at the politics of the sensibilities of current capitalism, which allow us to identify conglomerate actions that deny the normative content of neocolonial religion and are linked to collective ways of doing things: *Interstitial Practices* (IP) (Scribano, 2009, 2012b, 2013, 2017, 2019).

One way to understand these practices is to direct our gaze towards the unnoticed, porous, and occluded folds that exist in the daily experiences of the millions of expelled and discarded subjects from the Global South. We have observed that practices of "lived life" are made effective there, which as a power of surplus energy to predation, appropriate the open and indeterminate spaces of the capitalist structure, generating a "behavioural" axis that is located transversely with respect of the central vectors of a configuration of the politics of bodies and emotions. Therefore, they are neither orthodox or paradoxical practices, nor are they heterodox in the conceptual sense that Pierre Bourdieu gave them. Rather, they constitute inadvertent folds of the naturalized and naturalizing surface of how society accepts to distribute bodily energies and the correct ways of feeling. The practices to which we refer are updated and instantiated in the interstices, understanding these as the structural breaks through which the absences of a given system of social relations are made visible. These breaks are irregular spaces where the subjects build a set of relationships tending to weld the conflictual structure but with different and multiple tinges. These are welds that go through bodies and emotions, enhancing re-passionateness.

On the other hand, if we look back at collective actions from the 1950s to the present day, we can discover the existence of multiple experiences of struggles against the consequences of capitalist expansion, instantiated from different geographies and class positions/conditions. In the aforementioned framework, revolutionary, insubordinate, rebellious, and resistance practices emerged that marked the conflict of those years. However, our research carried out during the first decade of the new millennium has led us to identify a series of particular practices that coexist with the previous ones, but which have focused our attention: *Collective Interdictions* (CIs) (Scribano, 2012b).

In summary, some of the main features of CIs are the following:

a) These practices arise as a brake on violence, usurpation, and dispossession of what a collective(s) designates as common, and in this sense, many times they emerge accompanied by the diverse actors that carried out the previous social protests, such as social movements, NGOs, and "political actors" of the most diverse kinds.

b) They cannot be understood without taking up the previous conflictive networks, with different degrees of articulation and recovery of the memory and the memory of protection of threatened collective property. In this sense, CIs are an act of defence of meanings nested in particular geopolitics and geocultures, constituting one more twist of the Moebius bands that connect the constitution of "we" in its indeterminacy and contingency.

c)   In their antagonistic content, the CIs promote and retake an aesthetic of "so far, no more," of "this is too much." In other words, they constitute the action of "drawing a line in the sand" that marks the limits of dispossession, and they challenge patience/waiting as civic virtues of diminished democracies.

d)   They, in turn, constitute a temporal and spatial practice that involves the procedural redefinition of limits, margins, and edges. In other words, CIs are "time-space packages" that seek and elaborate meanings about what exists, delimiting and re-delimiting the common. Thus, if colonization practices involve actions to define territories based on limits arranged to guard the common goods and objects of dispossession, the CIs constitute a line to the advance of the colonial, becoming strong in what is collective in the usurped.

By knotting these elements together, CIs have emerged as (a) plural and contingent objections to the configurational ties of effective restraints, (b) as systematic oppositions to the deprivation of energy, and (c) as challenges to the existing forms of repression.

However, in addition to CI, it is also necessary to realize that in thousands of poor neighbourhoods and shanty towns, in innumerable urban or rural settlements, in countless communal, collective or public spaces, *Affirmation Experiences* (AEs) are performed every day (Scribano, 2015a). Said experiences are the set of shared feeling practices around interactions aimed at elaborating satisfiers in connection with collective/individual demands. The interactions that give rise to the aforementioned feeling practices can have multiple modalities, objectives, and degrees of institutionality collective, public and/or market. In the same direction, the demands with which the interactions are linked can be connected to lacks, needs, and/or desires. The various forms through which interactions and demands can be articulated depends on the different satisfiers that are elaborated, on their structures, and the meanings that are granted.

The AEs then take place within the framework of activities that range from dance classes, through cinemas and neighbourhood radio stations, to churches and political parties. There are thousands of interactions around dining rooms (children, the elderly, etc.), gyms, community micro-enterprises, collective services (water, electricity, etc.), housing construction, community rooms, squares, and so on.

Thus, there are ways of being in common that create an "us-others," an intersubjective connection according to shared ends and borders where the most evident (and valued) result is the affirmation of the common content of subjectivity. Feeling one, feeling that they are all one, developing a parenthesis in agonizing relationships, recognizing pairs, and feeling of modifications are, among several others, the possible contents of the feeling practices that constitute the aforementioned experiences.

They are practices where affective/cognitive association mechanisms are instantiated, which, at least partially, are associated with a form of collective identity and identification processes. Identifying oneself implies emotional investment, the inscription of energy and affective cognitive imitation, and this identification is one of the vault keys of affirmation as a practice of feeling. "Being-part-of," "building-a-space," recognizing, and being recognized become nodes of experiences where the radical intersubjectivity of subjective identity becomes a resource for affirming it. Thus, it is possible to perceive the appearance of a group of collective practices that we will call topologies of rejection.

As I have argued elsewhere (Scribano, 2015a), these imply the instantiation of countless interstitial practices, diverse experiences of affirmation, and various CIs. In this direction, the last three decades have left us contradictory and clear lessons on the social structuration processes concerning collective actions, political institutions, and social transformations. Neither the democratic actors that were renewed after the dictatorships and civil wars, nor the social movements to fight neoliberalism, nor the "social organizations" in the mentioned but not instantiated mobilization democracies, nor the sacrificial and spectacularized structures around charismatic leaders, none of these processes seem to have been efficient in the removal of the capitalist system as a regime of predation and expulsion. In the context of these three decades, the social structuration processes have been configured in the web of dizzying modifications that range from the massification of the Internet and social networks, through drastic redefinitions of conspicuous consumption, up to the multiple forms of daily violence (gender, ethnic, drug, etc.).

In this context, it is possible to identify areas of inadequate structures that we are going to understand as those irregular and unstable forms where a set of practices have settled that deny the reproduction processes of societies normalized in immediate enjoyment through consumption. These are zones that map the practices that are inappropriate and unattainable in the molds that are expected both from the normalization of society and from the appropriate "progressive political incorrectness." Understanding inadequacy as what deviates, comes out and contrasts with what is appropriate what is acceptable and accepted, as what is misplaced, out of place, and unexpected, a set of practices can then be identified with which it is possible to constitute nodes of figures/shapes as zones.

These zones are created despite the reproduction of a subject cut to the size of consumption, that are made despite the massiveness of some groups that enjoy spectacularization as a political space, that are stressed by the multiple ways of being silent. They are zones that in their dialectic weave the "duty," the "faults," and the inconsistencies that interstitial practices point to; being collective inter-dictions and experiences of affirmation that at the same time absorb and restructure their potentialities.

The interstitial practices that run through the day-to-day life involve family members, friends, neighbours, and relatives, they constitute those spaces between the individual/collective where accepted/acceptable sensibilities are denied. But these practices are not what sociology calls revolutionary or anti-system actions necessarily. CIs multiply and their density increases "proportionally" with the forces of expansion of capital on common goods, but they do not seek to change the system, but rather concentrate (with different levels of efficiency) on making it not "grow." The AEs are the privileged forms of collective actions in the contexts of the "world of no," racializing segregation and repression, but they are sustained with parts of self-blame and self-responsibility that do not question the mimetic consumption or the diminished humanitarianism of the neo religion-colonial.

It is precisely in this Moebius strip that implies the dialectic between interstitial practices, CIs, and AEs that have constituted an area of inadequacy that hatches/opens the possibilities of some other cartographies: *Topologies of Rejection.*

In this context, for us, the uncertain and contingent practices of rejection are associated with geometries of bodies and grammars of actions that co-constitute social topologies. They are topologies that feed on pregnancies that become acts of instantiation of old/new inheritances. Rejecting is also a consequence of refusing, of refusing to continue in reproduction, of ceasing to accept beyond the fact that only paths in unknown and labile morphologies can be identified.

The topologies of rejection are forms that make up contradictory fields of forces, morphologies of denial, and a Moebesian ribbon of denials. Saying "no," maintaining distance, and denying resignation, are practices that shape life lived in autonomy and are perhaps the key to future marches of the collective.

Neither interstitial practices, nor CIs, nor AEs alone are sufficient for an inaugural act of autonomy. We will have to strain the subtlety of the observation to capture the new situations where topologies of rejection are developed from the Moebesian tension between the three.

6.3.1 Messages, symptoms, and absences

The second vector that we would like to highlight here arises from a particular hermeneutic of the *rejections* instantiated in the transition to the new century, which proposes to interpret collective actions from the (a) messages, (b) symptoms, and (c) absences associated with them (Scribano, 2004).

a)   As messages, these actions manifest the state of the conflict resolution mechanisms, communicating in the direction of those edges where society can no longer set limits, and on those borders that have been crossed, demanding a work of systemic redefinition.

b)   The symptomatic, for its part, works by metaphorical transposition; of a sign, the meaning of a set of relations is interpreted to which that sign does not make direct reference but supposes them. In this direction, collective actions are signs of the processes of social production and reproduction in such a way that they make it possible to see what is happening within that process. They are symptoms in the sense that they allow visibility to what, by social logic, turns upside down or inverts, and to what this logic prevents immediate access.

c)   Finally, as absences, collective actions show the places where society has no cement, where it is not sutured. In this sense, the absences refer to those moments where the logic of the social cannot unite the natural ties between the agents that the relationships entail.

6.3.1.1 Messages. social networks: instrument or content

Depending on what the mobilizations that have opened this new century have been showing, we could critically return to Marshall McLuhan's affirmation that *the medium is the message*, to affirm provocatively: Social Networks (Facebook, Twitter, Instagram, among others) are the content/message, they are not mere vehicles or "means for." Indeed, if we review some of the movements popularized in recent decades, from the Arab Spring, through the movement *Occupy*, to the "Indignados," we can highlight that "what was left" as a message is "grab the megaphone," "participate among all," and "communicate through digital media," among others. It is not specifically "what they said" that we would highlight as "the messages," but precisely what is emphasized is "what they did." In this sense, the first learning about it results from highlighting the practice as the content of the *message itself*.

A logical *message* linked to collective action, as we conceive it here, is one that refers to the limits of systemic compatibility in such a way that it communicates us towards destructuring processes (Melucci, 1994). In this sense, the social networks placed in the context of these mobilizations are not only an instrument to convene, but are "in themselves," the very idea that is to be given, that is, the content of the action: the practice is the content.

In this direction, said content refers to thinking that the trans-class impact, although also classed, those networks have in the construction of sensibilities. The new century is also born on a 4.0 habitability, where vitality is placed in the sensible mediations of life 4.0: from ordering the pizza to joining a protest or demonstration, it is done on the Internet.

We thus highlight two elements to think about the collective action agenda: (a) in the first place, the contagion of emotions through social networks (Serrano-Puche, 2016; Ruiz Santos, 2015; Kramer et al., 2014; Marenko, 2010) lead us to qualify the link between these 4.0 practices and modes to reproduce sensibilities. And consequently, the logic of collective action, increasingly anchored in these 4.0 practices, will be affected in the way it is reproduced and circulated; (b) on the other hand, digital interfaces are postulated as superior to face-to-face relationships, as they provide possibilities for subjects to express themselves beyond words. In other words, the practice of protest 4.0 offers other expressive possibilities linked to the collective content of the actions: this includes the multiple resources that each platform makes available, from "emoticons" and "profile customizations," to the most diverse resources from which is constituted a "new visual literature" (Twine, 2016; Bresciani and Eppler, 2015).

6.3.1.2 Symptom. Depoliticized citizen

The symptomatic view of social conflict leads us to characterize an "ideal type"[3] of a subject who is the protagonist of the new century, and who can be described as a "depoliticized citizen," where "the political" has a strongly different meaning. This is why it was defined in the 20th century and is part of what has been called "compensatory consumption" (see Chapter 4) (Scribano, 2015b).

The context of the emergence of this subject is a society normalized by enjoyment through consumption. That is a political economy of morality and certain logics of mimetic consumption, which drives subjects and groups to identify with objects, in such a way that the rules of objects take shape and life, governing the life of the humans. But at the same time, logics of compensatory consumption are deployed, where the State has become a "state of subsidy."

This symptomatic gaze contributes at least three elements to reconstruct the agenda that summons us:

a) Firstly, the "depoliticized citizen" invests his time in consuming. A few decades ago, work constituted the structure that imprinted the rhythm of life of the subjects, while today – and increasingly – consumption is the organizer of every day: subjects live to consume, and therefore, there is no time for criticism, because consumption is something specific, immediate, and criticism involves "other temporalities" that connect "beyond" the spasmodic normalizing enjoyment. Thus, this depoliticized consumer has to do with this feeling of not having time to protest, to reflect, for autonomy.

b) Secondly, and within the framework of the aforementioned regarding the place of consumption in the normalization of societies, we must recognize that subjects can also "consume the experience" of "feeling indignant," of "feeling part of it," etc. Not only can they consume, but they can also participate in the experience, and even "without leaving their home" (through digital social networks). In this sense, it is necessary to ask ourselves about the effects that this logic of "banalization of the good" will have linked to collective actions.[4]

c) Finally, the symptom of the depoliticized citizen also leads us to think about the new centre of politicization of that subject. If the subjects can consume utopia through participation, if they have the possibility of consuming the will to transform, to build autonomous horizons of life, without a doubt one feature of the conflict will pass precisely through the dispute against consumption. That is to say, the rejection and resistance against the set of devices aimed at regulating sensations and the mechanisms of support, linked to structuring consumption. But also, those against the set of institutions and productive structures that prepare subjects for consumption. In this sense, if this century that began with the idea that the dispute is in one, the new century seems to be built on the verification of the dispute "with oneself."

6.3.1.3 Absence. Naturalization of violence

Absence, as we will understand it here, constitutes a strong generator of presences. And in this sense, what the collective actions "reveal" is the very "ferocity" of the "depoliticized citizen." Thus, as we have been describing, although the "ideal type" of the protagonist of the collective actions of this new century leads us to portray a citizen willing to be subsidized, this portrait is complemented by the lines that characterize a compulsive consumer. But this compulsion, softened in the friendly times and spaces of the market, denotes (in its absence) the naturalization of violence that completes the image of the subject's ferocity.

Thus, daily life is traversed by corporations, the true "structure of marketing" of society, where the key to "social reproduction" is to move further and further away from contact with violence. But on the other hand, violence is also around us every day: from the growth figures that draw an ascending scene of urban violence, to the unheard cries that outline the millions of "other abjects" who are daily thrown into repressive mechanisms, which for their part do not stop developing, privatizing, and specializing.

This panorama related to the reading of *absence* refers us to three central questions to ask about the agenda of collective action:

a) What happens with a world where it is said that "everything is fine" and, in contrast, the daily experience refers us to deaths, urban violence, gender violence, slavery, etc.? It is precisely here where we must emphasize that we are facing a system that denies its consequences, that is, violence and even cruelty. The other is an object to destroy: by definition, the consumer is someone who destroys, therefore one of the features of collective actions may be linked to care as a denial of said destructive logic.

b) Linked to the above, and even, in the context of what has been said regarding "the territorial," there seem to be more and more territories where the monopoly of "physical" violence by the State is questioned. Thus, it would be expected that the logic of collective action is structured in terms of making violence that is naturalized visible, even beyond State structures, or the "claim for rights."

c) Finally, this tendency to "naturalization," to "act as if it did not exist," that is, this normalization of violence, is undoubtedly linked to the complex mechanisms of fear management. Thus, the collective actions of the current century could be established as platforms for presenting processes related to the "emotionalization of violence," that is, to make explicit certain mediations between a particular regime of sensibilities and particular situations of expropriation.

6.3.1.4 Virtues and practices

The third vector that we want to trace based on the agenda for thinking about actions is linked to the possibility of recovering "virtues"[5] (this is discussed in detail in Chapter 7) between the gaps of some of the most widespread ideological practices today. Thus, the commodification of life as a feature of the daily life of the capitalist world has disabled the discussion of virtue as a mere aftertaste of "morality," showing (as concealment) that such disabling is nothing more than deliberate forgetting of the concrete guidelines that the political economy of morality imposes, weaving a set of practices of feeling as ideological practices.

The virtuous, the virtue, and the virtues are not thematized, in a game of non-actions from which it is forbidden to talk about them as a device of forgetting and stigmatization that consecrates them to the world of active phantoms of the possible foundations of life.

In this direction, we return to the possibility of rethinking the virtues as Moebesian tensions and the processual obverse of compensatory consumption, as a bearable mechanism associated with the processes of predation, dispossession, and expulsion of capitalism in our societies. The compensation produced by the State today is through the consumption and amnesia about the virtues as parameters of life in common. It is in this context that living in common demands rethinking the virtues in consonance with, at least, three of the elements that are the object of the regimes of oblivion: the power of the other as co-living and equal; caring for oneself as a condition for the possibility of listening; and the removal of hatred and envy as a parameter of action and feeling practice.

Following this line of analysis, we can account for a paradox that has been instituted in the emotional ecology elaborated in the first few years of the 21st century: the more the discursive is appealed to as a support for politics, the fewer the opportunities exist to listen (to) the subjects. In a web of discursive slogans (and overvaluation of the media) the subjects of flesh and blood, the feeling practice of these people and the institutions associated with them, are spectators of desires and talks in which they do not participate and that do not belong to them.

Marketing, as a scientific discipline that studies the production and management of emotions, has mainstreamed the policy-making sense for both the market and the State. In line with the above, the enormous effort of the capitalist structure to produce normalization entails structures of silencing the opportunities to have their own voice and throat on the part of the millions of subjects "exposed" to the regimes of forgetfulness and the politics of memory.

The elision of the possibilities of saying through the spectacularization of society (market/State) restricts the potentialities of the practices of feeling that are collectively constructed in the thousands of experiences of "common talk" that exist. It is from this perspective that a possible point on the agenda linked to the collective action of this century is linked to the possibility of generating collective conditions for the expression and autonomous management of desires and voices. In other words, we would expect the emergence of a series of subjects/collectives that retake desires where mimetic consumption, resignation, and compensatory consumption policies "have left" them, elaborating a collective reconstruction on the connections/disconnections between needs/desires/demands, which implies a "cut" and a rejection of the normalizations that immediate enjoyment imposes as a rule of interaction. In this same direction, we would also expect the emergence of a series of praxis linked to common speech, where the plots between identities/narratives/inequalities can be put into a crisis beyond the breakdown of the "marketing" slogans of the policy that became "sale of emotions."

Finally, and associated with the above, from a collective treatment of wishes and common speech, where the self-absorption in the desires built by is minimized *marketing* and instantiated in immediate enjoyment, "other" social forms of listening could also emerge.

These might promote the production of collective sayings structured in the expression of their own voices and re-appropriation of the throats. It is from the spaces of power that give autonomy to common speech that the construction of another reality can be reinvented, where others cease to be inert objects of individualistic jouissance (*sensu* Marx), and reknot the tasks of a co-living with all living beings.

Notes

1 At present there are several ways of understanding collective emotions that are not of the same view that we want to give to emotional ecology here, but that must be mentioned as close "antecedents" CFR (Flam and King, 2015; Von Scheve and Salmela, 2014; Garcia and Rimé, 2019; and Scribano and Lisdero, 2017).
2 For a review of the literature on emotions, collective actions and conflict CFR (Scribano 2019).
3 We warn that by "ideal type" we are not thinking of a Weberian construction of it. Instead, we propose it in a sense more attached to the resonances that this structuring process (normalization through enjoyment through consumption) implies as a desired "model" of the subject.
4 The banalization of the good, as a logic that configures one of the traits linked to normalized societies, has been described in Chapter 1 with two other complementary logics are also put in tension: the "politics of perversion" and the "logic of waste". In this sense, the questions raised about the connections between "banalization of the good" and collective action in the context of the new century could be extended around the complementary trends alluded to.
5 Some of the notions presented here have been developed in Scribano (2014). Regarding our perspective on virtues as social practices, it draws on various sources, among which we can cite: Bhaskar (1987); Marcuse (2001 [1968]); and Melucci (1991).

References

Bresciani, S. and Eppler, M.J. (2015) "The pitfalls of visual representations: A review and classification of common error made while designing and interpreting visualizations", *SAGE Open*, October-December. doi: 10.1177/2158244015611451.

Bhaskar, R.A. (1987) *Scientific Realism and Human Emancipation*. London: Verso.

Flam, H. and King, D. (2005) *Emotions and Social Movements*. London: Routledge.

Garcia, D., and Rimé, B. (2019) Collective Emotions and Social Resilience in the Digital Traces After a Terrorist Attack. *Psychological Science*. 2019 Apr; 30 (4): 617–628.

Kramer, A., Guillory, J. E., and Hancock, J. (2014) "Experimental evidence of massive-scale emotional contagion through social networks", *PNAS*, 111, (29). www.pnas" www.pnas.org/cgi/doi/10.1073/pnas.1320040111

Marcuse, H. (2001 [1968]) "Beyond one-dimensional man". In *Collected Papers of Herbert Marcuse*. Volume 2. Editado por Douglas Kellner. London UK: Routledge.

Marenko, B. (2010) "Contagious Affectivity. The management of emotions in late capitalist design", *Contagious Affectivity: The Management of Emotions in Late Capitalist Design. 6th Swiss Design Network Conference*, Basel, Switzerland. ISBN 9783952366219.

Martineau, H. (1838) *How to Observe. Morals and Manners*. London: Charles Knight and co.

Melucci, A. (1991) *Il Giocodell'io*. Milán Italy: Feltrinelli.

Melucci, A. (1994) ¿Qué hay de nuevo en los nuevos movimientos sociales? In Laraña, E. and Gusfield, J., *Los Nuevos Movimientos Sociales. De la Ideología a la Identidad.* Madrid: Centro de Investigaciones Sociológicas.

Ruiz Santos, P. (2015) "¿Qué sabemos sobre el contagio emocional? Definición, evolución, neurobiología y su relación con la psicoterapia", *Cuadernos de Neuropsicología/ Panamerican Journal of Neuropsychology*, 9, núm. 3, diciembre. pp. 15–24.

Scribano, A. (2004) "Conflicto y estructuración social: Una propuesta para su análisis." In Zeballos, E., Tavares Dos Santos, J. V. and Salinas Figueredo, D. *América Latina: Hacia Una Nueva Alternativa de Desarrollo.* Arequipa: Universidad de San Agustín.

Scribano, A. (2009) "Sociología de la felicidad: el gasto festivo como práctica intersticial", *Yuyaykusun*. N° 2, Lima, Perú: Departamento Académico de Humanidades de la Universidad Ricardo Palma, pp. 173–189.

Scribano, A. (2012a) *Teorías sociales del Sur: Una mirada post-independentista.* Argentina: ESEditora-Universitas.

Scribano, A. (2012b) "Interdicciones colectivas, violencia y movimientos sociales hoy", *Revista Actuel Marx/Intervenciones*, N°13, segundo semestre, Chile: Santiago.

Scribano, A. (2013) "La religión neo-colonial como la forma actual de la economía política de la moral", *De Prácticas y Discursos. Cuadernos de Ciencias Sociales*, Año 2 no. 2. CES-UNNE. Resistencia: Universidad Nacional del Nordeste - Centro de Estudios Sociales.

Scribano, A. (2014) "A look at some acts of violence and silenced repressions: Evictions in Argentina", *Research on Humanities and Social Sciences*, 4, (5), pp. 68–79.

Scribano, A. (2015a) "Acción colectiva y conflicto social en contexto de normalización", *ONTEAIKEN*, N° 20 - noviembre 2015.

Scribano, A. (2015b) *¡Disfrútalo! Una aproximación a la Economía Política de la moral desde el consumo.* Buenos Aires: El Aleph.

Scribano, A. (2017) "Amor y acción colectiva: una mirada desde las prácticas intersticiales en Argentina", *Aposta. Revista de Ciencias Sociales*, 74, pp. 241–280.

Scribano, A. (2019) *Love as a Collective Action: Latin America, Emotions and Interstitial Practices.* London UK: Routledge.

Scribano, A., and Lisdero, P. (2017). Saqueos en la Argentina: algunas pistas para su comprensión a partir de los episodios de Córdoba - 2013. *Caderno CRH*, 30(80), 333–351.

Serrano-Puche, J. (2016) "Internet y emociones: Nuevas tendencias en un campo de investigación emergente", *Comunicar*, XXIV(46), pp. 19–26.

Twine, F.W. (2016) "Visual Sociology in a Discipline of Words: Racial Literacy, Visual Literacy and Qualitative Research Methods", *Sociology*, 50, (5). https://doi.org/10.1177/0038038516649339.

Von Scheve, C. and Salmela, M. (2014) *Collective Emotions: Perspectives from Psychology, Philosophy, and Sociology.* Oxford: Oxford University Press.

7 Values

7.1 Introduction

As a consequence of the analysis of the colonization of the inner planet, structures, collective actions and an exploration of the dialectic of the person, it is necessary to return here to what was left pending in Chapter 1 regarding the connections between ethics and aesthetics in terms of the politics of sensibilities accompanying the diagnoses offered. As anticipated in Chapter 1, currently this can be seen in the emergence of a neo-religion of helplessness. This (institutional) policy ought to create the new religion of neocolonially dependent countries to replace the old trinity of "industrial religion" based on unlimited production, absolute freedom, and unrestricted happiness, with the trinity of the expelled composed of mimetic consumption, diminished humanism, and resignation. It is a religion whose liturgy is the construction of social fantasy, where dreams play a central role as the kingdom of heaven on earth, and the frustration of sociodicy's role of narrating and making present and acceptable the phantasmic hells of the past felt like a present continuous. Mimetic consumption implies a set of transversely connected practices that are developed in the form of the absorption of properties of an object as an act of appropriation, shaped by fantasizing on the qualities of the subjective carrier of such action. Diminished humanism is a relationship of the suture of absences registered in a person carried out by one or more other person that leave intact the processes that cause such absences. Resignation is a way to make bodily (like embodiment) the non-future (un-future) of expectations and desires of a person constructed as a result of the acceptance of the limits imposed by their material conditions of existence as a closed totality.

In the context of societies normalized in immediate enjoyment through consumption, the emergence of three characteristics of the aforementioned social processes can be verified, namely: the structuring of a Logic of Waste (LoW), the elaboration of Politics of Perversion (PoP), and the Banalization of Good (BoG). In this sense: (a) the structure of social relations in a waste society is understandable by the normalization of society in and through the sacrificial spectacularity which implies, among other components, the ritualization of a "putting oneself in the hands of tomorrow" as a fantasy of redemption, the connections between waste/classification as metamorphosis of the inequality system, and practices of feeling molded from the disposable, the discarded, and the discarding as interaction; (b) the PoP consists of making lies, manipulation, and fictionalization a desirable "state of affairs" as a central strategy for managing emotions; they are political because their modulation and execution must be thought of the patterns of feeling that are elaborated in the tensions between sociabilities and experientialities, managing sensations, managing emotions, precisely with the aim of directing the social cement that is found/elaborated/arises between ethics, morals, and aesthetics; and (c) The BoG has at least six basic characteristics that are threaded into two pairs of elliptical and dialectically arranged triads:

(1) fetish-dogma-heteronomy and (2) epic-gesture-narration. These triads are registered and displayed in an irregular space qualified by four complementary nodes of resignation, as a component of the neocolonial religion: fake contention/dependence/romanticism /miserabilism. The two triads are, in short, moments of the same helical movement where each of its moments comes in tension with the other and with the passing at different times through the same place, but in a different "state of affairs." Thus fetish-dogma-heteronomy and epic-gesture-narration integrate the features of the BoG and are located on an inscription surface constituted/drawn within the framework of four points that interact geometrically: fake contention/dependence/romanticism /miserabilism.

Social structuration processes since the beginning of the 21st century have been characterized in the Global South, among many vectors, by mutations in the contents, limits, and volumes of social policies as mechanisms of conflict elision, transformations in the forms and management of work, and the changes in the use and commercialization of space.

Along with neocolonial religion and LoW, the elaboration of PoP, and the BoG, it is possible to find in the scopic regime of Society 4.0 another axis of dialogue and dispute over the values and political economy of the morality of current capitalism. Seeing–feeling begins with touching–looking at yourself. The image is an intersubjective production that acquires the characteristic of instantiated practice at the moment that the production made for the viewer is captured. While making an image I touch the surfaces of some devices that I need to look at to see myself feeling what I want to know and make known. Today seeing is touching-feeling what is seen. The fingertip makes contact with the screen(s), the glass receives the pressure of a decision-making, and when sliding it navigates the options menu that the previous selection enabled. When we see a photo, we are touching it, at times almost imperceptibly, but most of the time with that moment of monitoring that prevents mistakes: the unwanted like, the wrong upload, and the wrong stalking. In this mode of interaction, human beings are elaborating a grammar of vision more here as a code than the word. The instaimage[1] is a proposal to live an experience from the immersion in a scene that transits the sensibility and sensoriality of making images known with our hands. It is the image-makers who, determined to communicate sensorialities, are redefining the experience of feeling (oneself) in and with (the) image. They are actors in daily productions where the features of the new "visual literacy" are played out, they are persons that with the "camera-at-hand" always privilege the production of the ICI effect (immersion, connectivity, and intensity). Instaimage production is guided by capturing not photographing, looking to capture, not a photo, trying to transmit an experience, not an object in a massive and radically self-produced way, it is a synthesis of a scopic regime that from the old produces "new" consequences. Although every image seeks to transmit experiences, the instaimage is based on this quality of "portraying" and uses it as a starting point. Are modifications in our scopic regime also modifications in our value system? If an ethic corresponds to every aesthetic transformation and a politics corresponds to it, perhaps the answer is yes.

Now, the fact that these practices and not others are used to "make-understand" in which paths empathy with the posted image is sought, opens the way for us to ask ourselves about them as interstitial practices. Neither apocalyptic nor digitally integrated into the era of touch and at the end of the Anthropocene, we must accept that love and family, as well as beauty and creativity, continue to be means of expressing our human way of inhabiting the planet. The acceptance that we are "feeling thinking" beings (*sensu* Fals Borda) lead us to wonder about our "video-touching" condition as producers of sensibilities that allow us to know/feel the world. The challenge for the social sciences of societies normalized in immediate enjoyment through consumption in the context of the 4.0 revolution remains how to link/unlink science and politics.

It is in the game of the dialectic of the person, as reconstructed in the previous chapter, that the possibility of denying the claim of totality the political economy of morality appears. This possibility implies a reflection on some of the central axes of some interactions created from listening, dialogue, and hope as its opposite.

7.2 Beyond vertigo: silence as a starting point

To begin the end of this book, it is good to look in a classic for a footprint to discuss the possibilities of remaking the world in a common way among all living beings on the planet:

> If we suppose man as a man and his relationship with the world as a human relationship, we can only change a love for love, trust for trust, etc. If you want to enjoy art to be an artistically educated man; If you want to exercise influence on another man, you have to be a man who acts on others in a stimulating and exciting way. Each one of the relations with the man – and with nature – must be a determined externalization of the individual real life that corresponds to the object of the will. If you love without awakening love, that is, if your love, like love, does not produce reciprocal love, if through a vital exteriorization as a loving man you do not become a loved man, your love is impotent, a misfortune.
>
> (Marx, 1974 [1844]: 183)

As we have already considered, we live in a society normalized in immediate enjoyment through consumption that has reconstructed the **politics of the senses** (seeing, touching, tasting, smelling, hearing) and **politics of sensibilities** (organization of life, information to order preferences, and values to the parameters for the management of time/space), and in this way reformulated the **political economy of morality** (BoG, the LoW, and the PoP) (Scribano, 2017a, 2018).

The continuities and discontinuities between the 20th and 21st centuries that this painting represents impose a review of the ways of knowing and intervening available to the social sciences, among many reasons, because we subjects have transformed ourselves with the same speed and depth as the structural features referred to.

It is a cognitive-affective priority of the social sciences of the 21st century to reflect and intervene on the current politics of listening and silence.[2] The proposal for dialogue that we want to leave raised as a "prologue" to our practices is structured on the helical configuration between listening/silence/listening as a platform to break/complement the visual metaphor of knowing: moving on from the question: What gaze do you have? What listening do you display?

Silence is the flow where sounds, noises, speech, music, etc., are expressed. The practices of expressing are a break/continuity of a constitutive absence in the original configuration of the individual/agent/actor/subject/person.

Before following the footsteps of silence, let us return to the action of hearing: "*...the word" hearing "comes from the Latin auditio and means" action of hearing. " Its lexical components are audire (to hear), plus the suffix -tion (action and effect)."*[3]

Silence is preparation for perception, since the vibrations captured by the ear "resonate" (Pérez-Colman, 2015) in the person with everything social that is in them and what is involved in this act of "mapping the world." The conscious and active exercise of hearing transforms hearing into attentive, reflective, and transformative listening.

> The word" listen "comes from the Latin auscultate (to apply the ear) formed of auris (ear). The second element of the compound is of disputed etymology, but has been related to the Indo-European root klei- (to bow). This root is the one that gives in Latin the verb clino (to incline) and its compounds decline and incline.[4]

From the individual body (Scribano, 2012a), "putting the ear" (putting the body), passing through the subjective body in both, inclines and skew/tilt and reaching the social body of the inclination as reclining/bending the auscultation is an "attending with the body." Silence is an activity crossed by a body/emotion that ties the distances between individual and subject, preparing knowledge of the other and of the others approaching the ear by initiating an active hearing that allows you to hear. The voices, the steps, the vibrations, and the sounds of the near and far are transformed into astrolabes to guide the journey through possible worlds. In the journey, the connection between knowing how to listen and being silent is displayed as the beginning of the dialogue with the context and others.

Social scientists are, in a special way, attentive listeners to those expressivities that draw us to the social world not yet said in the academy.

Silence is the opening act of the urgency of the other/others to break the absence of words as complicity and the silencing as a violence of everyone's listening. Hearing/silence/listening tensions is a propaedeutic as a beginning and as a preparation for knowing, knowing (oneself), and knowing (us) in society: sociology is a listening practice that has its starting point in silence.

Billions of words, images, sounds, screams, and moans populate a world of images in and through touching (Scribano, 2015), demands silence as a practice of rupture and reflective appropriation of the thousands and millions of people who hide in said pornography of seeing.

The question "What are you listening?" implies (a) keeping silent as a break with anxiety about the reason for self-referred speech; (b) making explicit how it is heard, how one body approaches another to understand what it has to say from its sociabilities, experiences, and sensibilities; and (c) clarifying what is expressive in the silence of the listener. Silence is the starting point of dialogue as a matrix of knowledge that has become personal interaction.

7.3 Voice, listen, and society starting a critical dialogue

The social sciences are a way to make suppressed dialogues emerge, they are a path of reconstruction of marginalized dialogues and are emphatically the product of a dialogue between people who inhabit the social world and know it. For a long time, we have been insisting that Latin American social sciences should be thought from silence as an opportunity to listen to everyone, to listen to each other, and to listen to the sound of the world (Scribano, 2001, 2019). Social sciences as a dialogue between human beings, as a dialogue where all living beings on the planet participate, as a dialogue where all words and all grammars participate, is a task of reconstruction of the multiple knowledges that nest in suppressed dialogues, unheard voices, and hidden conversations.

We have emphasized that silence is the point where speech begins, we have emphasized that silence is the call that we make for the other to speak. Here we want to return precisely to the structure of the process that is born after the silence, that is, the voice. There is a deep relationship between the voice, the word, and the grammar of action. According to Amanda Weidman

> ...The voice is a set of sound, material and literary practices formed by cultural and historically specific moments and a category invoked in the discourse on personal agency, communication and representation, and political power.
>
> (2014: 38)

Much has been written about the semiotics of the word, much has been written about the grammar of action, but we still have to think and rethink of a look from the social sciences on the voice, on the tone, on the tonality, and the "gesture phonic." For this reason, in homage to the book presented here, we will offer a very brief sketch of the importance of the voice for the social sciences in general, and for sociology in particular. We going to try to synthesize the importance of this sound, which serves for the coordination of action and therefore for the structuring of the social.

At the beginning of the last decade, Paul Prior argued

> ...voices are often expressly represented as personal and individualistic or socially as a system of discourse. Based on sociohistorical theory (particularly Voloshinov and Batkhtin), in this article, I defend a third view in which the voice is simultaneously personal and social because the discourse is understood as fundamentally historical, situated and indexical.
>
> (Prior, 2001: 55)

In a different context, but close to Prior's intention, here we want to argue that there are different types of voices: personal voices, collective voices, and institutional voices. The individual voices are those that we produce as subjects, as actors, and as agents, they are the vehicle of the interrelation that occurs with the word. They are not only the support of the word, but also the instrument of communication, each person has a voice. Also, the voices are collective, they are summative, they are articulations of words that support words that carry meanings, that aim to narrate, but have a tone and a sound that is characteristic of it. For this reason, collective practices produce voices that can be understood as an expression of the body that constitute that set of individual practices that we call collective, but that are characterized by the surplus that this gives to personal voices. Finally, there are institutional voices, voices of sociability, voices of norms and rules that we hear, and which are produced in the context of the institutionalization of life. That is why they are in politics and are called spokespersons, there are also communicators who speak in the name of the institution, there is always someone in charge of producing the institutional voice, a person, or a group of people. The relationship between personal voices, collective voices, and institutional voices produces social speech; your speech not only differs but are also related, your speech is in tension, and your speech is articulated to produce the encounter, conflict, and distance.

So, each person, each collective, and each institution has its sounds, has its ways of emitting sounds that are characteristic of it. Through voice, we recognize persons, through voice collectives they make their identity known, and through voice institutions they communicate and build a particular sociability. There are voices in the conflict, in the scream, in the onomatopoeia; all are diverse expressions of how the voice is taking shape, which is the fruit of man's contact with reality through breathing. The intensity, the cadence, and the colour of the voice represent the action, they put the action back in front of the others. The intensity implies a volume that is valued according to the interest of the emphasis of the locution, of the need for the word to be expressed in the structure of the narrative. The cadence is geoculturally determined; there are accents, ways of saying, there is a relationship between pronunciation and you in the word. The voice also has its colour, it transmits different shades between accent and regularity. There is a geoculture and a geopolitics for the tones of the voice.

Thus, it is possible to understand how the voice then connects with emotions, with identities, and with the gesture. The voice (so) carries the emotions, the emotions are constructed/expressed primarily through the voice. Sadness has a particular type of voice, joy is expressed with a singular voice, and anger is manifested with a special voice, to give only three examples. Emotions are tied to the voice that challenges us in its intensity, in its colour.

The voice is a producer of emotions, when human beings listen to the voices, we produce emotional practices, hence, we often refer to voices of alarm, to voices of fear, and to voices of war. From this perspective the voice is an emotional incarnation, the voice lends an organ to the expressiveness of emotion. There is a very carnal sense of emotion in the voice, that is why in singing the timbre of the voice identifies not only the logic of the particularity of the person, but also of what contextualizes the emotional expressiveness of that sound. Within this framework, it becomes understandable why disputes and articulation unfold daily to "manage" the voices that express emotions, to lower their volume, and to make what they evoke inaudibly. The voice carries, channels, and expresses emotions.

That is why there is a direct relationship between voices and identities, as individuals we have our voice associated with the organic structure of our phonation apparatus articulated epigenetically to our insertion in the collective and family history. Each human being has their voice, but it is radically dependent on context, it is always a situated practice. The voice has to do with the individual body which is built phylogenetically and ontogenetically. The voice is that which, when listening to it, identifies the one who produces it, the voice is the first step to recognition, the voice is an organic particularity of individual identity. When I am listening, my voice is always the interaction that defines me. But there is also a voice that has to do with the persistence of traits that make the person who produces it a different individual, a different subject, and a different actor; there is an association between our voices and our dramatization of actions.

For this reason, there is also a relationship between voice and gesture; there may be a gesture without a voice, but not vice versa. When hearing voices, even if the subject is not seen, it is associated with a particular gesture. It is associated with fatigue, it is associated with joy, it is associated with exaltation, and it is even associated with silence. A voice puts in tension the mimetic expression; in the facial expression, the variation in facial features that a voice produces is an element of the interaction. In this same direction, intonation, the voice helps the word to incarnate, the word to take shape. The voice is related to the emphasis of the face, with the dialectic of the masks, with the naturalness of the face; the voice allows coordination, the voice allows articulating the ways of presenting the face. Studying the voice is an approach to understanding the gesture. A gesture of autonomy is embodied in the independent production of a particular voice.

This tension between your identity, emotions, and gesture must be placed in relation to what we have already mentioned about the personal voice, the collective voice, and the institutional voice, and in relation to what we had preannounced about the voice being a vehicle for dialogue, conflict, and fighting. That is why much of what is in collective actions in protests, in social movements, is the expression of collective voices that want to give a voice to those who do not have a voice. That is why many times the personal voices represent other personal voices that are crossed out, that are suspended, and that are erased. For this reason, institutional voices are a gesture of recognition of rejection of indifference concerning subjects, individuals, and actors.

As Jenny R. Lawy puts it, "a voice is heard when listeners participate in the dominant forms of speech and action through which the voice becomes acceptable, readable and audible" (2017: 194). That is why, in this context, it makes sense to return to the voice as not only that point of convergence, but also of the distance between shout and word.

Because there are precisely voices of resistance, there are also voices of dispute. After all, some voices embody collective interdictions, because there are voices of insurrection. The voice accompanies the cry to transform into a word that in terms of demand will be made public as a word spoken by all. The history of collective voices is the histories of the structuring of the social dispute for the right to say yes also with a different identity. After the silence the voice comes, the voice produces the plot of the real, supporting the word by giving meaning to the cry.

The systematic study of the proximity and distance between collective, personal and institutional voice allows us to obtain clues about the processes of autonomy and dependence, of submission and emancipation. As Frantz Fanon wrote:

> The colonized, therefore, discovers that their life, their breathing, his heartbeat is the same as the settler's. Discover that a colonist's skin is not worth more than an indigenous skin. It must be said that this discovery introduces an essential upheaval in the world. All the new and revolutionary security of the colonized follows from this. If indeed, my life has the same weight as that of the settler, his gaze no longer glares at me, it no longer immobilizes me, **his voice does not petrify me**. I no longer get excited in his presence.
>
> (Fanon, 1963: 22, emphasis mine)

These connections between voices, silences, and identities allow the configuration of a set of practices of encounter and dialogue where the five sides of the geometry of the person are allowed to take different paths back to those behind in the current modalities of coloniality.

7.4 Hopes, virtues, and life in common

For a long time, there has been research and writing about the current situation of capitalism, coloniality,[5] and the interstitial practices associated with it. In these contexts, it has been proposed to repair love, happiness, and reciprocity as "forgotten" practices, underlining their character of denial of neocolonial religion.[6] Closing this book on the colonization of the inner planet, it is necessary to re-postulate an analysis in this regard.

The purpose of this section is to synthesize some notes on the possibilities of a "life-another-in common." Taking and/or making "notes" has three functions, or it can be understood from three perspectives: (a) it is a long narrative; it is written to record where it has passed, to indicate the ports that it has touched; (b) they are texts with some features that I am going to fix on paper so as not to forget; signs of a crossing; and (c) the notes are open diagrams; sketches that materialize what has been "thought out loud."

What is written below is an invitation for us to think about the future, as a practice denied by the neocolonial religion whose bases are mimetic consumption, diminished humanism , and resignation. It is a denial that conjures up any act of positioning oneself from and in the future as a useless and meaningless practice. The future is something that cannot be thought fundamentally for one reason: if we think about it, maybe we will.

Thinking/saying/doing the future is a disruptive practice insofar as it specifically indicates the hiatuses of social domination that is presented as a pure "that's the way things are." *Thinking about* the future demystifies the past as destiny; *saying* the future removes the sacralization of the present; *to do* the future is to present in the "here-now" one life and another as autonomy.

Breaking with all romantic, miserabilistic, and avant-garde presumption, breaking the seduction of speaking for others, and locating ourselves more here than in academic practices as priestly actions of "those who know most," this section intends to share some notes that allow us to reconstruct various (and plural) intineraries, journeys, and sketches of life in common. For the proposed task, the following argumentative chain has been drawn: (a) to trace the recovery of the "virtues" between the gaps of some of the most widespread ideological practices today, (b) look critically at the connections between wishes and common speech, and (c) rediscuss the notion of hope.

7.4.1 The phantom of virtues

The dictionary of the Royal Spanish Academy defines "virtue," in its first meaning, as an "activity or force of things to produce or cause their effects,",[7] while in the fifth it states: "integrity of spirit and goodness of life;" and it is in the interlude of the powers between one and the other that we believe it is important to rethink life in common.[8]

The commodification of life as a feature of the daily life of the capitalist world has disabled the discussion of virtue as a mere remnant of "morality," showing (as concealment) that said disabling is nothing more than the deliberate forgetting of the concrete guidelines that the political economy of morality imposes by weaving a set of practices of feeling as ideological practices. The virtuous and the virtues are not thematized in a game of nonactions from which it is forbidden to talk about them as a device of forgetting and stigmatization that consecrates them to the world of active phantoms of the possible foundations of life.

Retaking the problematic of the virtues is one of the chapters of the dispute of the implicit power in the consolidation of the neocolonial religion at present; it is to operate a deconstruction of the fallacies that clothe the virtues of the robes of substantialism, ontologisms, and dogmatism, and it implies removing the regimes of forgetting from their totalizing effect as one of the main components of the devices for regulating sensations.

Our proposal is to initiate the task of re-dialoguing the virtues, taking them as Moebesian tensions and the procedural opposite of compensatory consumption as a supportability mechanism associated with the processes of predation, dispossession, and expulsion of capitalism in our societies.[9] Compensatory consumption is a process that is inscribed between the folds of the current accumulation regimes, state compensation systems and the expansion of market logic. Compensatory consumption is today the main public policy aimed at reinstalling the effectiveness of developmental modernity as the cement of colonial societies. Far from being characterized as neo-developmentalists, the current forms of the State in the Global South must be thought of as and from their own accumulation regime. Transversally, adolescent capitalism (as opposed to its supposed senility) has structured a set of political regimes that make the expansion of consumption its main policy aimed at stabilization and conflict elision. The metamorphosis of the State has produced sociabilities, experiences, and sensibilities that, like the great companies and world corporations, are designed to the extent of the production, management, and reproduction of sensations. The classic passkey of the State as the mediator of the conflict that consisted in the elaboration of wage goods (education, health, tourism, etc.) has shifted to its capacity to generate a "type" of consumption that fulfils triple functions: (a) naturalizes predation, (b) expands the reproductive capacity of the various fractions of the capitalist classes in power, and (c) grants the necessary means for the consecration of immediate enjoyment as the axis of daily life. To summarize the argument: the compensation produced by the State today is through the consumption and amnesia of the virtues as parameters of life in common.

It is in this context that living in common demands rethinking of the virtues, following at least three of the elements that are the object of the regimes of oblivion: (a) the power of the other as co-living and equal, (b) caring for oneself as a condition for the possibility of listening, and (c) the removal of hatred and envy as parameters of action and practices of feeling.

a)   As co-inhabitants of a time-space we share a biography (s), we are living beings-with-others, and we inhabit a "here-now" since we participate in a legacy (s) that marks our condition as pairs, peers, and equals with others. The structure of the regimes of forgetting seeks to make this co-experience invisible, the manipulation of the past, and the elaboration of an "always like this" as a result of a politics of memory. The task of reliving the virtue (s) as a disruption of forgetting implies the recovery of being with others in a world whose forms start from the equivalence/difference of living as close. The other appears as the contingency that in its radical freedom makes possible the experience of a we-others as an indeterminate basis for the exercise of living. Because we co-inhabit, we co-live, and it is in the dialectic recognition/hetero-recognition of these proximities that autonomy allows us to display virtues.

b)   One of the effects of compensatory consumption and immediate enjoyment is the loss of listening as the axis of interaction from the difference, overcoming the inequality of dispossession as a homogenizing relational practice. Living beings with neighbours in the exercise of their memory are built from socialized and extended self-care practices as a bridge of mutual attention. Taking care of yourself begins to take care of the other in a triple sense: to pay attention while listening from the throats of those who interact, to notice the continuities/discontinuities between demands/desires/needs of those who express themselves, and to concentrate on the being of each one of those who "do in common." Taking care of yourself is the starting point of taking care of your partner; taking care of yourself is giving yourself the opportunity of autonomy against the traps of consumption as the only means of contact with others; caring for oneself is to grant/accept/sustain subjectivity in its character as a radically intersubjective result.

c) For the co-experience with/of the other to be possible from acts of listening made possible by intersubjective care, it is necessary to remove hatred and envy as features of daily interactions. Hatred is the aversion to the living that eliminates intersubjective possibilities; it is the practice of erasing the qualities of a different; it is the impulse to carry out extermination. Envy is the practice of feeling rooted in animosity about the possibilities of enjoyment of the other. Hate and envy are the characteristics of a society centred on the possession, dispossession, and violence of inequality. Go from mimetic consumption to co-experience with the living where no one can be put in the place of the object, go from immediate enjoyment to collective happiness practices, go from having to "being-being-with-others" in a world shared, dissolves the power of hatred and envy as social forms of a political economy of morals made flesh. In this context, battling against the phantasmatic situation of virtues in a social world governed by neocolonial religion implies re-gazing at the place of desires and a common speech.

7.4.2 *Listening together: an outline about the desires[10] and common speech[11]*

A paradox has been instituted in the plots of feeling elaborated in the early years of the 21st century: the more appeal is made to the discursive as support for politics, the fewer opportunities of listening (they) have the subjects. In a web of discursive slogans (and overvaluation of the media) the persons of flesh and blood, the practices of the feeling of these people, and the institutions associated with them, are spectators of desires and talks in which they do not participate and that do not belong to them.

Marketing, as a scientific discipline that studies the production and management of emotions, has mainstreamed the development of the politics of the senses for both the market and the State. That is to say, market/State at present is based on the management of desires through two mechanisms: (a) economic policies oriented towards consumption and (b) compensatory consumption policies as the main axis of "social peace."

In line with the above, the enormous effort of the capitalist structure to produce adequate levels of ideological surplus value[12] entails structures of silencing the opportunities to have their own voice and throat on the part of the millions of persons "exposed" to the regimes of oblivion and the politics of memory. The elision of the possibilities of saying through the spectacularization of society (market/State) restricts the potentialities of the practices of feeling that are collectively constructed in the thousands of experiences of "common talk" existing in the world.

It is from this perspective that the collective conditions of expression and autonomous management of desires and voices become the object of all critical/ utopian thought. The central idea is to retake desires where mimetic consumption, resignation, and compensatory consumption policies "have left" them. Elaborating a collective reconstruction on the connections/disconnections between needs/desires/demands implies a "cut" and a rejection of the normalizations that immediate enjoyment imposes as a rule of interaction. Constructing praxis of common speech, where the plots between identities/narratives/inequalities can be put into crisis by leaving the iron cage, implies the mere solidarity gestures and the dismantling of the marketing slogans of the politics that have become a "sale of emotions." Thus, thinking of common speech between equals as a praxis of autonomy, where illuminist temptations and criticisms coagulated in a revolutionary "as if" are erased, begins with a renewal of acts of listening.

From a collective treatment of wishes and common speech, where the self-absorption in the desires constructed by marketing and instantiated in immediate enjoyment is minimized, the social forms of listening can be recovered that enhance the production of collective sayings structured in the expression of subjects' own voices and a re-appropriation of throats. It is from the spaces of power that give autonomy to common speech that the construction of another reality can be reinvented where others cease to be inert objects of individualistic jouissance (*sensu* Marx) and reknot the tasks of a co-living with all living things. These are common words that, placing the phantasmatic and fanciful elaboration of desires at the centre of their preoccupations, break with the fatality of the "no way out" and dislocate the appearance of a closed totality of resignation. It is common talk that returns to the vitality of the past as a battering ram against the repetitive return of horror and suffering through memory as the first consequence of the break with immediate enjoyment as amnesia of the collective. These are common words that, by removing the identity between consuming – enjoying – wanting from its sacred halo, find in listening to a being-being-among-equals that allows retaking the future as a now.

7.4.3 Hope is now

Hope[13] implies a set of anticipatory practices for the future. Hope is detached from the logic of patience and waiting as civic virtues promoted/built from the neocolonial religion and as instantiations of the nucleus of the political economy of the moral of the current state of the planetary expansion of capital. The hopeful attitude is a rejection of the avoidance of conflict and the pornography of dispossession; it implies a break with respect to the claim of the validity of resignation as a closed and immutable totality; it consists of the dispute of the fanciful qualities of immediate enjoyment and it is a practice that redefines inhabiting the world with others as equals.

Hope is plotted on a day-to-day basis from the senses that interstitial practices provide as daily denials of the absoluteness of the truth regime on the inevitability of capitalism. Hope is co-constituted in the ellipsis of a life lived as a reciprocal sharing that reinhabits tomorrow. Hope is the result of the guarantees given by millions of testimonies about remembering as the first act of a memory policy. Hope is woven into and from the networks of a being-being for the fruit that understands itself as a Moebesian process of construction of another world. Hope is a way of inhabiting our times/spaces differently. Hope is a practice that challenges us to reinvent the future, anchored in the redefinition of yesterday, today, and tomorrow. Hope is today, as is the future.

The future is now: because every day we elaborate what we will be in each choice of coexistence; because it means the "presentification" of what we have been disputing, the conditionalities of what was "now;" and because it houses what is in it of common practices and collective interdictions. The future involves the forms from which we recompose the webs of our practices of feeling, updating them from a hopeful attitude as a collective practice. To think/do/say the future is to desecrate what is of horrific surplus dispossession in the present. To paint the world in the colours of hope is to build the future.

Today, 20 years into the 21st century, the colonization of the inner planet is being systematically completed and in the face of this, a radical critique of the available knowledge is needed to scrutinize the procedures, paths, and vehicles of colonization. It is within this framework that the development of social sciences that can perform this task from a trans-disciplinary perspective becomes imperative.

After these notes, this writing sketched as the record a voyage, making evident the features of a still incomplete log, there is still much to do/say/think. The social sciences, in commitment to utopian thought, have a long way to go. Weaving hope with the possibilities of caring for ourselves as close beings who co-inhabit a time/space is today one of the central challenges of these social sciences. Taking back the virtues and rebuilding utopias away from prejudices and encouraged by the daily force of interstitial practices is a central task for the university and the scientific system of the planet.

The task requires some social sciences commitments, involved with the criticism of some rationalities centred on the bureaucratization of knowing/doing/feeling and registering in the dispute for the recovery of the potentialities of virtues, building alternative models for knowing (us). In this direction, an alternative model of knowledge is above all a collective task, where knowledge is a lifestyle that contributes to the democratization of the colonization of the "natural," social, and internal worlds. It is a knowledge that bets on the potentialities of a radical democracy of intimacy, autonomy in life decisions and human emancipation.

In this context, perhaps, the direct connection of hope with love and politics can be made as Martin Luther King Jr. argued in the last century:

> Power properly understood is nothing but the ability to achieve purpose. It is the strength required to bring about social, political, and economic change... And one of the great problems of history is that the concepts of love and power have usually been contrasted as opposites–polar opposites–so that love is identified with the resignation of power, and power with the denial of love. Now we've got to get this thing right. What [we need to realize is] that power without love is reckless and abusive, and love without power is sentimental and anemic... It is precisely this collision of immoral power with powerless morality which constitutes the major crisis of our time.
>
> (Luther King, 1967)

Thus, social sciences that think/say/make the future will already be the first node of a network that takes one/us/others from one knowledge to another in the service of collective happiness.

Looking into each other's eyes and dissolving the monochromaticity of threat that capitalism imposes on us maybe the first step to scandalous communitarian love.

Notes

1 For the notion of instaimage CFR Scribano, (2017b).
2 For a vision different from the one expressed here on silence CFR Le Breton, (2006).
3 http://etimologias.dechile.net/?audicio.n
4 http://etimologias.dechile.net/?escuchar
5 On our vision regarding the connections between capitalism and coloniality, *CFR* Scribano, (2012b).
6 For a recent look at our view on neo-colonial religion and interstitial practices, *CFR* Scribano, (2013a).
7 http://lema.rae.es/drae/?val=virtud
8 Our perspective on virtues as social practices draws on various sources, including Bhaskar, (1987), Marcuse, (2001 [1968]), and Melucci, (1991).
9 In connection with compensation, consumption and "development" see chapter 4 and regarding our theming of enjoyment, consumption, and spectacle, *CFR* Scribano, (2013b).
10 Regarding the complex notion of "desire," a set of clarifications would be needed that for reasons of space I can present here only as an outline of indication CFR Marcuse, (2001 [1968]).
11 For a close but also diverse look at the issue CFR Milner, (2011).
12 Regarding the development of the notion of ideological surplus value CFR Silva, (1984 [1970]) and Scribano, (2010).
13 We return here, in a heterodox way, to what was stated in Bloch, (1996).

References

Bhaskar, R. (1987) *Scientific Realism and Human Emancipation*. London: Verso.
Bloch, E. (1996) *The Principle of Hope*. Volume 1. UK: Basil Blackwell.
Fanon, F. (1963) *Los Condenados de la tierra*. México DF: FCE.
Lawy, J.R. (2017) "Theorizing voice: Performativity, politics and listening", *Anthropological Theory*, 17, (2), pp. 192–215. https://doi.org/10.1177/1463499617713138.
Le Breton, D. (2006) *Du Silence*. Paris: Editions Metai lie.
Luther King, M. (1967) *Where Do We Go From Here?* The Martin Luther King, Jr. Research and Education Institute, Strandford Univertiy. https://kinginstitute.stanford.edu/where-do-we-go-here
Marcuse, H. (2001 [1968]) "The Movement in a New Era of Repression." In Kellner, D. (Ed.) *Collected Papers of Herbert Marcuse*, Volume 3. UK: Routledge UK.
Marx, K. (1974 [1844]) *Manuscritos: Economía y Filosofía*. Madrid: Alianza.
Melucci, A. (1991) *Il Gioco dell'io*. Milano: Feltrinelli.
Milner, J.C. (2011) *Pour une politique des êtres parlants(For a politics of speaking beings)*. Paris: Verdier.
Pérez-Colman, C. (2015) "The sound field and the ear of sociology: From the sonorous doxa to the sociological ear, or the theoretical-analytical foundations for the study of a sound life", *Methaodos. Revista de Ciencias Sociales*, 3, (1), pp. 107–120.
Prior, P. (2001) "Voices in text, mind and society: Sociohistoric accounts of discourse acquisition and use", *Journal of Second Language Writing*, 10, pp. 55–81.

Scribano, A. (2001) "Investigación cualitativa y textualidad", *Cinta de Moebio. Revista de Epistemología de Ciencias Sociales*, (11). Available at: https://cintademoebio.uchile.cl/index.php/CDM/article/view/26302/27602

Scribano, A. (2010) "TESIS 1: Colonia, knowledge (s) and social theories of the South", *Onteaiken Bulletin on Collective Action Practices and Studies*, No. 10, Year 5, pp. 1–22. http://onteaiken.com.ar/boletin-10

Scribano, A. (2012a) "Sociología de los cuerpos/emociones", *Revista Latinoamericana de Estudios sobre Cuerpos, Emociones y Sociedad - RELACES*. N°10. Año 4. diciembre 2012-marzo de 2013. Córdoba. pp. 93–113. http://www.relaces.com.ar/index.php/relaces/article/view/224

Scribano, A. (2012b) *Social Theories of the South: A post-independence look*. Buenos Aires: ESEditora. Córdoba: Universitas - University Scientific Publishing House.

Scribano, A. (2013a) "Neo-colonial religion as the current form of the political economy of morality", *Practice and Discourses Magazine*, year 1, N° 2, pp. 1–20. http://ces.unne.edu.ar/revista2/pdf/Scribano-Dossier.pdf

Scribano, A. (2013b) "A conceptual approach to the moral of enjoyment: Normalization, consumption and spectacle", *Brazilian Journal of Sociology of Emoção*. 12, (36), pp. 738–751. Paraiba, Brazil-ISSN 1676–8965.

Scribano, A. (2015) "Beginning of the XXI century and social sciences: A possible puzzle", *Polis* [Online], 41, Published on September 20. http://polis.revues.org/11005; doi: 10.4000/polis.11005

Scribano, A. (2017a) *Normalization, Enjoyment and Bodies/Emotions. Argentine Sensibilities*. NY: Nova Science Publishers.

Scribano, A. (2017b), "Instaimagen: Look by touching to feel" (Dossier "As for reasons and as emotions das images"), *RBSE Revista Brasileira de Sociologia da Emoção*, 16, 47, pp. 45–55.

Scribano, A. (2018) *Politics and Emotions*. Houston: Studium Press llc.

Scribano, A. (2019) "A modo de prólogo. El silencio como punto de partida". In Cervio, A. and D'heres, V. (Comp.) *Sensibilidades y experiencias: acentos, miradas y recorridos desde los estudios sociales de los cuerpos/emociones*, pp. 11–13. Buenos Aires: Estudios Sociológicos.

Silva, L. (1984 [1970]) *The Ideological Surplus Value*. Caracas: Editions of the Library, Central University of Venezuela.

Weidman, A. (2014) "Anthropology and voice", *Annual Review of Anthropology*, 43, 37–51.

Index

Page numbers in **bold** indicate tables and page numbers followed by n indicate notes.

Printed in the United States
by Baker & Taylor Publisher Services

Printed in the United States
by Baker & Taylor Publisher Services